Contact

Arca
4 avenue de Coeuilly
94500 Champigny-sur-Marne
France
+33 (0)9 5338 4412
www.arca-home.com

Airwood

Composition

Wood and rubber

Properties

Decorative, amazing, almost 'alive', and customised motifs possible

Applications

Interior architecture and signage

Technical specifications

Thickness: 2 mm or thicker

Skilled cabinetmakers, inventors of this principle, accompany us to a world where objects could be of variable geometry. Their patented technology in effect allows the material to be inflated (or sucked in). Wood, but also ceramics, hard plastics, metals, leathers or corks for example, can thus come to life before your eyes. This technique allows motifs, writing, light, a function, a form, an indication to be made to appear and disappear. The principle is to chase or chisel the material so as to fragment it and release its tensions. The material thus worked is then glued to a rubber membrane before fixing the whole onto a sealed substrate. A pump fed by a compressor – or even your breath – is then going to regularly deform the material, as if it is breathing.

1

2

3

1 *Trac-T*, retro sucked up seat / Ebony, stainless steel / Length 450 mm / Width 450 mm / Height 450 mm

2 *Plink*, decorative panel, vertical rain / Oak / Length 1200 mm / Depth 30 mm / Height 1200 mm

3 *Épure*, air purifying panelling / Oak / Depth 100 mm / Height 2500 mm

10

Contact
Annemette Beck
Lindehusvej 5
5750 Ringe
Denmark
+45 (0)6264 1564
info@annemette-beck.dk
www.annemette-beck.dk

Annemette Beck

Composition
Monofil, viscose raffia, Lurex, Nylon, paper yarn, metal threads (silver, stainless steel, copper, brass, etc.), bike inner tubes, crystals, coated polyester, natural fibres, sisal, etc.

Properties
Decorative and effect with light play

Applications
Room dividers, rugs, sun shades, decorative panels, curtains, upholstery fabrics, etc.

Technical specifications
Fabrication limits: solely within the bounds of possibility

Annemette Beck is a Danish textile designer who experiments with materials and structures in her weaves. Her sophisticated work, which may be hand or machine made, explores the qualities of paper, rubber threads or metal threads – silver, copper, bronze or stainless steel – or even recycled bike tyres, wool, cotton or Lurex, for example. Whether she is creating rugs, wall coverings, curtains or furnishing fabrics, her philosophy remains one of simplicity. She gets down to essentials, but her creations are always relevant and elegant, never lacking in imagination or a touch of fantasy.

See notions: **Nylon** p.227

12

Contact
Ekobe
Rua Pe. Teòfanes A.Barros 80
57033-000 Riacho Doce
Maceió - Alagoas
Brasil
+55 (0)82 3355 1128
ekobe@ekobebrasil.com
www.ekobebrasil.com

AOBA Line

Composition
Coconut shell

Properties
Decorative and strong

Applications
Floor, wall and ceiling coverings

Technical specifications
· Dimensions: 200x200mm, 300x300mm and 700x700mm
· Wide choice of colours and finishes available

These panels are made using small, square pieces of coconut shell, the arrangement of which generates different surface finishes. One can thus choose to show the exterior or the interior of the shell, i.e. the pieces can be convex or concave. Finally, this product offers a chance to 'recycle' coconut shell, which is generally discarded once the nut has been removed.

1, 2 & 3 The AOBA line is used for wall and floor coverings.

1

3

2

Contact
JSP
1443 E. 12 Mile Road
Crown Office Village, Building J
Madison Heights, MI, 48071
United States
www.jsp.com

ARPRO

Composition
Expanded polypropylene (EPP)

Properties
Light, resistant to impact, chemical attack and heat

Applications
Automobile parts, packaging, household electrical goods, furnishings, etc.

Technical specifications
· Different surface finishes and colours are available
· Possible densities from 0.20 to 2.50 g/cm^3

16

Compared to expanded polystyrene (EPS), ARPRO is more flexible, more resistant to successive impacts, to chemical products and to heat. ARPRO beads are injected into a mould, where they agglomerate under the action of steam, heat and pressure. A solid moulded object can have curves, corners and complex shapes.

1

3

2

4

1 A new, easy-to-install car park protection system. Using ARPRO significantly improves vehicle and building fabric protection. The 'Park Sentry' product provides improved protection for pedestrians, cars and car park pillars.

2 Lightweight, modular and durable, the Q-Couch made from ARPRO has endless possibilities. Today you need a set of single-seaters, tomorrow a long couch to fit all your friends in a row? No problem, you can arrange and re-arrange the Q-Couch as you need it.

3 ARPRO can be made in any colour you like. In addition to black and white there are four bright colours in the ARPRO sample pack.

4 The featherweight LYTA chair (11 kg) is an ergonomically-shaped armchair, made from ARPRO. Easy to handle and move around, and completely recyclable. The LYTA structure is emission-free and allergy-friendly, as no metals or glues are used. Covers and foam cushions can be removed for cleaning.

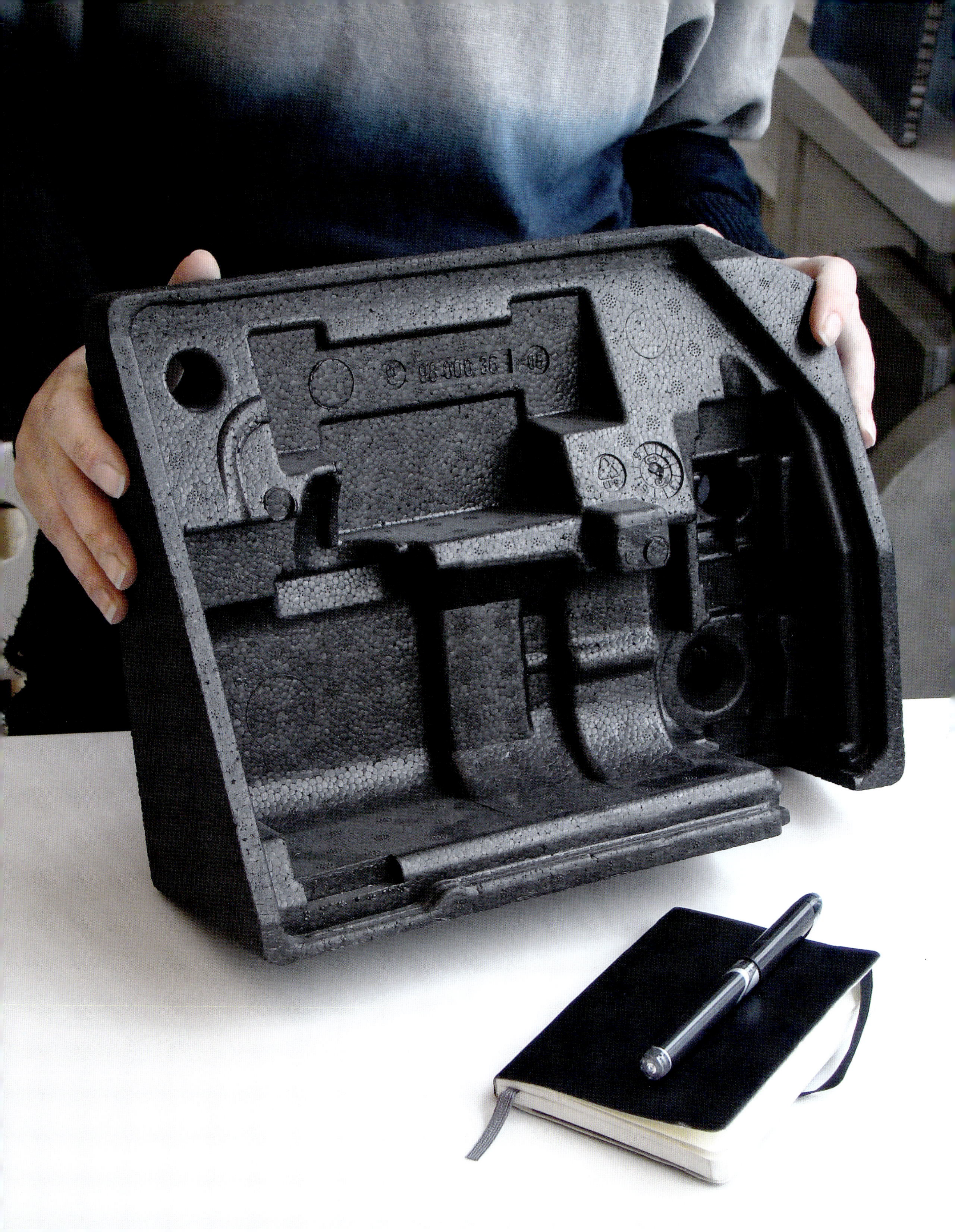

Contact

Arty Beton
17 avenue des Sources
84000 Avignon
France
+33 (0)6 8318 3050
artybeton@aol.com
www.artybeton.fr

Arty

1

Composition

High-performance concrete: loaded with siliceous limestone, binders and additives

Properties

Decorative, delicacy of patterns, high-precision detail and relief

Applications

Signage, interior architecture, furnishings, etc.

Technical specifications

- Dimensions: 5x5cm to 125x200cm
- Density: 2 to 2.1 g/cm^3
- Elastic deflection: 10 MPa
- Compression: 50 MPa
- Modulus of elasticity: 28,000 MPa
- Laying method: adhesive / mortar
- Can have integral optical fibres
- Can have a large range of colours. The concrete can be coloured, polished or given a patina
- Panels can also be used for wall facings, tiles or paving slabs

These high-performance concrete panels are poured into moulds produced by digital imaging, opening the way to all sorts of fantasies. Concrete becomes decorative, its fine texture making it suitable for any pattern or detailed relief. Several collections are available: Velvet Concrete inspired by nature, Optique Concrete with graphical and contemporary emphasis, Baroque Concrete with delicate tracery. The company is also open to creation of new effects.

See notions: **Fibre-reinforced concretes** p.221

18

2

1 & 3 The patterns of the Velvet Concrete Collection are borrowed from nature. The matt and gloss structure interacts with the light.

2 The Optic Concrete Collection is very graphic and contemporary.

3

Contact
Azuna
114 Old York Road
Jenkintown, PA, 19046
United States
+1 215 925 9001
info@azuna.net
www.azuna.net

Azuna™ 3D

Composition
Polypropylene

Properties
Decorative, 3D-optical effect, light and thin

Applications
Packaging, design and merchandising

Technical specifications
Unlimited number of possible images

4-colour prints are made in 3D on a recyclable polypropylene substrate, giving an impression of great depth. Images can have three to five perceived layers. This is not a lenticular technique.

1 This notebook cover features a skull with a 3D optical effect because of the Azuna™ technology.

2 Bicycle number plates featuring a 3D optical effect because of the Azuna™ technology.

1

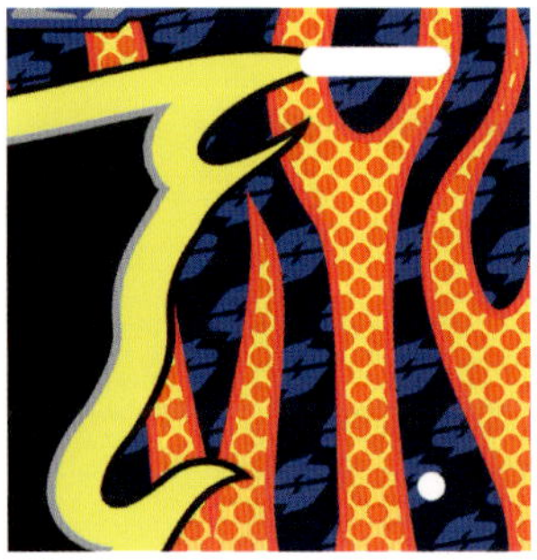

2

20

Contact
B-ton design
425 rue de Grignon
78450 Chavenay
France
+33 (0)1 3054 3234
www.btondesign.com

B-ton Scintillant®

Composition
Concrete and glass beads

Properties
Decorative, induces light play, resistant to compression, waterproof and wear-resistant

Applications
Architecture, interior and exterior

Technical specifications
Appearance is equally attractive in natural or artificial light

This material is made using BSI Ceracem® (ultra-high performance fibre-reinforced concrete) with glass beads in the surface. Light plays on them and the appearance seems to change with the viewing angle.
It picks up light and changes colour like a chameleon.

See notions: **Fibre-reinforced concretes** p.221

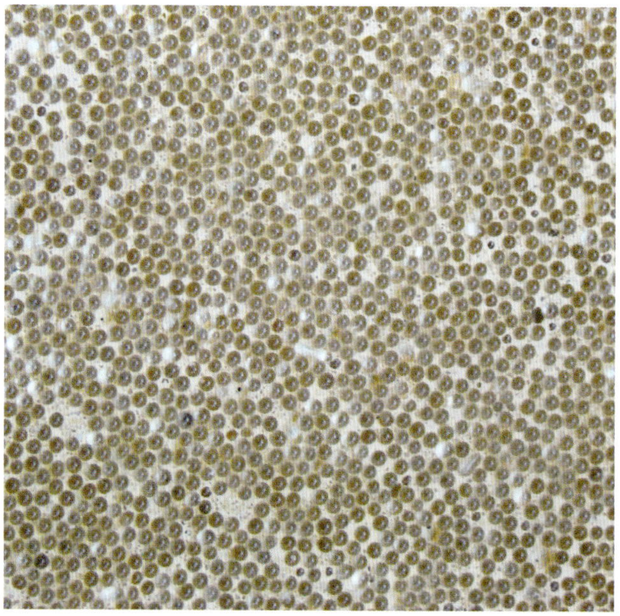

22

Contact
FIBandCO
Habitation Rivière la Manche
97224 Ducos
Martinique
+596 (0)596 7545 72
contact@fibandco.com
www.fibandco.com

Banana Core

Composition
Banana pseudo-trunk

Properties
Decorative, different natural colours and biodegradable

Applications
Veneer, wallpaper, paper, fashion accessories, interior decoration and luggage

Technical specifications
· Maximum dimensions: 130 x 300 cm
· Thickness: 0.1, 0.2 or 0.3 mm
· Natural colours
· Colours, patterns and thicknesses can be produced to order
· 100% natural
· 100% biodegradable

Banana plants are grown in the tropics (10 million hectares – 25 million acres worldwide). The raw material for this product is 100% banana pseudo-trunk. The pseudo-trunk is cut into a sheet, variously known as banana veneer, paper or papyrus. In this application, assembly of the decorative veneer sheet and the colouring process are accomplished by environmentally-friendly methods.

1

2

1 & 2 Conceived and created by Carine Hayot Architecture, this console Zuri presents a simple, modern and subtle design revealing a natural motif.

Contact
Schoeller Textile
Bahnhofstrasse 17
9475 Sevelen
Switzerland
+41 (0)81 786 0800
info@c-change.ch
www.c-change.ch

C-Change™

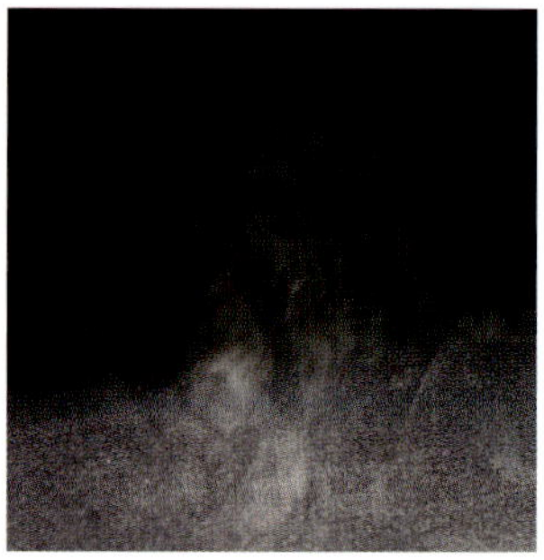

1

Composition
Polymer

Properties
Reacts to changing temperatures and activities, high breathability, creates a pleasant body climate in a wind and waterproof garment or shoe, and conforms to the bluesign® standard

Applications
Fashion industry: wear for sports, outdoors, leisure, travel and work

Technical specifications
Can be processed for non-elastic to highly-elastic membrane solutions

No more shivering or overheating. This membrane reacts to temperature changes and activity, offers a very high level of permeability to sweat vapour, is water-proof and wind-proof. The wearer's body is therefore kept in an optimum body climate.
The technology used in this membrane is inspired by observations in nature: the opening or closure of elements according to climatic conditions, for example the scales on pine cones. At high temperature or during intense activity, the structure of the membrane opens when the body humidity increases. Excess heat then escapes outwards. After a decrease in skin humidity during cold or inactive periods, the structure recloses to retain body heat inside the garment.

See notions: **Biomimetics** p.217

1 Excess body heat and moisture can escape to the outside air.

2 A sample of C-Change™ fabric.

3 The textile is water-proof.

2

3

Contact
Cubo Coverings
Industriestrasse 4
84323 Massing
Germany
+43 (0)87 2491 1230
info@cubo-covers.com
www.cubo-coverings.com

C4 Mosaik Böden

Composition
Wood, stag horn and epoxy resin

Properties
Decorative and several possible colours

Applications
Wall and floor coverings

Technical specifications
· Dimensions: 500x700mm
· Thickness: 19mm
· Available in several forms: random or regular pattern
· Backgrounds in all Pantone® colours are available on request

These panels consist of a mosaic of round slices of genuine horn, embedded in a translucent resin. The whole assembly is mounted on a wooden base. Another style employs the same principle, with a mosaic of wood slices of branches in oak or larch. Several background colours are available.

1

2

32

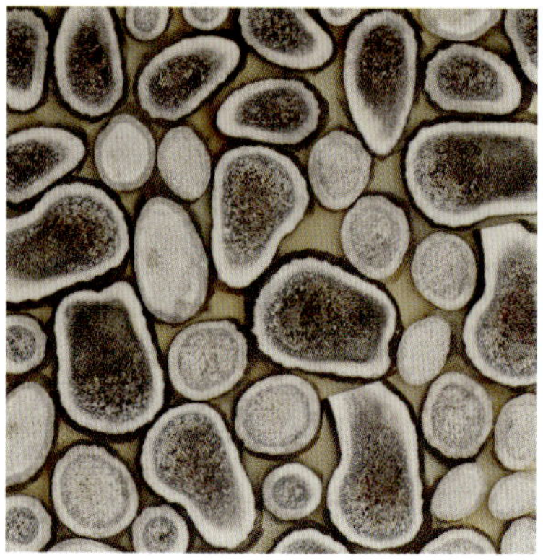
3

1 *Cubo c4, mosaik larch*

2 *Cubo c3, gummipflaster*

3 *Cubo c4, hartshorn*

Contact
Axente
1 allée d'Effiat
Le Parc de l'Evénement - Bât. H
91160 Longjumeau
France
+33 (0)1 6910 5070
contact@axente.fr
www.axente.fr

Chromapowerpix 200

Composition
LED, diffusion domes and lenses

Properties
High programmable light power and multiple colour effects

Applications
Interior architecture, events, etc.

Technical specifications
· 250 mm grid
· Primary mounting is by an optional suspension cable system with hanging bars
· Camlok Spacer Straps and Omega Clamps are also available to allow Chromapowerpix 200 to be deployed in many ways with ease

Chromapowerpix 200 is an advanced LED fixture, featuring 12 high-power pixels on a 250 mm grid, with digital-multiplex (DMX) control. Each pixel contains five three watt tri-colour devices giving an extremely high LED power rating of 15 W/pixel. A total of 180 W of LED tri-colour devices are built into every Chromapowerpix 200 and are driven by Pulsar's smooth, continuous-light, flicker-free and step-free drive electronics, all powered by a robust 200 W power supply, with world-wide compatibility. Chromapowerpix 200 is supplied complete with wide-angle diffusion plates and white overlay to allow choice of a black or white front panel. Optional diffusion domes and lens plates in a variety of beam angles are available for total beam-control flexibility.

Contact
Gray Concrete
Unit 1 West Shore Business Centre
Edinburgh - EH5 1QT
United Kingdom
+44 (0)79 7330 9146
info@grayconcrete.co.uk
www.grayconcrete.co.uk

Concrete Pillow

Composition
Concrete reinforced with glass fibre

Properties
Decorative, strong, durable and impermeable

Applications
Suitable for interior and exterior use, like furniture, wall covering, etc.

Technical specifications
Slender items: 8 mm minimum

Concrete offers almost infinite possibilities are far as the choice of texture, colour, shape and scale are concerned. Gray Concrete does not have standard products or techniques. Instead, it works directly in collaboration with designers or architects to create unique effects with concrete. Any shape or finish, from matt to polished, can be achieved – limited only by the bounds of imagination. By using concrete reinforced with glass fibre – which allows very thin pieces to be produced – light products are obtained, perfectly adapted for wall coverings for example, or for other projects where concrete does not always come to mind as the most appropriate material.

See notions: **Fibre-reinforced concretes** p.221

36

Contact
Villani Leonello
Via Limitese 158
50059 Spicchio Vinci (Fi)
Italy
+39 05 7197 9552
info@villanileonello.com
www.villanileonello.com

1 & 2 A sofa and lampshades made of natural cork fabric.

3 A versatile and resilient material used in upholstery.

© Courtesy of Villani Leonello

Cork Fabrics

Composition
Cork

Properties
Flexible and thin

Applications
Home furnishings, luggage, footwear, bookbinding, etc.

Technical specifications
· Dimension of rolls: 140x500cm
· Waterproof: Teflon treatment
· Resistant to abrasion and 30°C washing
· Can have fire-resistance treatment
· Different references and colours available
· Different substrates: cotton, Nylon, polyurethane textiles, etc.

Very thin layers of natural cork, laminated with different textile substrates. The composite material has all the characteristic softness and appearance of cork while being stronger and more resistant to wear. Armchairs and sofas, footwear, bags or walls can be trimmed with this velvety material.

See notions: **Nylon** p.227

38

1

2

3

Contact

Didier Versavel
67 rue de Fontenay
94300 Vincennes
France
+33 (0)1 4328 4490
contact@didier-versavel.com
www.cuirdabeille.com

Cuir d'abeille

This 'honeycomb' panel, of 100% leather, consists of a three-dimensional, latticework lambskin structure sewn to thin layers of transparent Irish sheep or goat parchment.

Composition
Leather and parchment

Properties
Decorative, translucent, resistant to compression, thick and light

Applications
Room layouts, leather goods, fashion industry, furniture, light fittings, displays, etc.

Technical specifications
· Dimensions: 250 x 250 mm, 400 x 400 mm and other sizes on request
· Different colours and printing possible

40

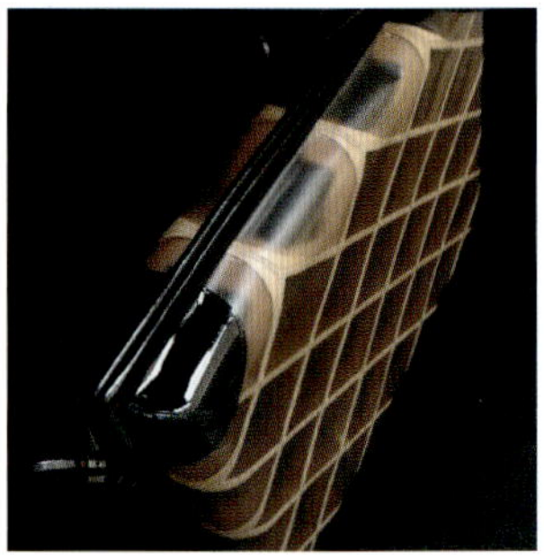

1

2

1 & 2 Designer Didier Versavel made a prototype of his designer briefcase in collaboration with upholsterer Guillaume Bernard de Bayser in 2009. The briefcase is made from a 100% leather structure with a translucent skin.

© Antoine Duhamel

Contact

Abinitio
Kronängsvägen 3
513 70 Borgstena
Sweden
+46 706 500 588
info@abinitio.se
www.abinitio.se

Cullus Trevira CS

Composition

Polyester

Properties

Acoustic insulation, 3D and decorative

Applications

Interior acoustic insulation

Technical specifications

· Standard widths: 600 and 800 mm, and on demand, up to 1200 mm
· 2 options: Cullus MOLTO weight: 1.5 kg/m², absorbent class A and C or Cullus MINI weight: 900 g/m², absorbent class D
· 100% Trevira CS polyester
· Fire resistant
· Machine wash at 30°C
· Colours: white, silver, grey, blue and yellow

Cullus, aimed at public spaces, is a unique, acoustically ultra-absorbent textile which combines functionality, flexibility and aesthetic qualities. Results obtained in use show that the acoustic absorption is proportional to the surface area of material used, whatever its form.
It is a knitted material which can be formed into 3D objects and thus arranged in rooms which need sound absorption.
Abinitio was started by Ulla Eson Bodin, Folke Sandvik and Helga Aiff in 2007. The material originated from a project at the Swedish School of Textiles, which focussed on 'smart' textiles.

1 & 2 At premises where it is not possible to have wall-mount absorbers, Globen is mounted to the ceiling to create a favourable acoustic environment. Globen is a globe-shaped wire structure covered with 5.5 m² Cullus MOLTO, which can be removed and washed.

42

1

2

Contact
Clearel Paris
150 rue Legendre
75017 Paris
France
+33 (0)1 4809 3874
contact@clearelparis.fr
www.clearelparis.fr

Dalle virtu (Virtual Tile)

Composition
PMMA (Polymethyl Methacrylate)

Properties
Very thin, and for an astonishing optical effect

Applications
Wall decoration, point-of-sale publicity, signage, artwork, etc.

Technical specifications
· Dimensions: 280x280 mm
· Virtual sphere: 220 mm diameter, over a 110° viewing angle
· Colours: black, red, blue, green, orange and mauve. Possibility of other, customised, colours for more than 50 units (chosen from a colour chart)

44

The tile is a micro-structured, acrylic plate which 'generates' one or two virtual spheres which appear to move according to the viewing angle. The colour can vary according to the material placed behind the optical-effects plate (self-coloured PMMA, or colour filter). One can also imagine putting an image instead of this coloured background to create an additional effect. The virtual tile offers the possibility for positioning an object at the centre of the virtual sphere with a thin fixing kit. There is then a sensation of an object levitating in a bubble.
There is also another type where two spheres can be visible: one of 180 mm and one of 90 mm.

1

1, 2 & 3 Examples of what the virtual tiles look like when moulded to the wall.

2

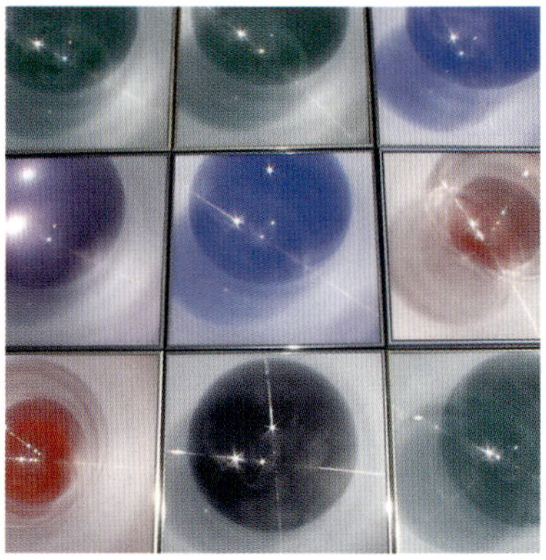

3

Contact
Turrini David Lange
Lieu-dit Keriaquel
56620 Pont-Scorff
France
contact@davidlange.fr
www.davidlange.fr

Daquacryl

Composition
PMMA (acrylic) and MDF or battens

Properties
Decorative, lustrous, smooth and durable

Applications
Interior architecture, decoration, furnishing, point-of-sale advertising, signage, etc.

Technical specifications
· Numerous opaque and metallised colours available
· Density: 1190 kg/m^3
· PMMA thickness: 3 mm
· High-lustre, very smooth surface: 98 gloss
· Excellent performance with respect to atmospheric agents
· Very good resistance to UV: no yellowing
· Excellent resistance to abrasion
· Can be cut, drilled, engraved, polished, glued, thermoformed and accepts screen printing
· Relief effects can be envisaged (quilting, for example)

This surface treatment technique consists of thermoforming a sheet of PMMA onto MDF panels or battens. The effect obtained is comparable to a lustrous lacquer but this material is solid and allows creation of furniture with a superb, lasting lustre. In effect, this self-coloured, PMMA does not lose its colour, shine and polish over time. This 'veneering' system allows an excellent finish to be obtained on edges so that it is even possible to cover the whole panel to give the illusion of a solid lacquered block.

1

1, 2 & 3 Sideboards from the David Lange Neptune collection.

4 The E PURE dresser, with its 2 doors in Daquacryl and bright chromed metal feet, was designed by Sacha Lakic and manufactured by David Lange

2

46

3

4

Contact
Decovision
Zone Industrielle
Rue Marc Seguin
63600 Ambert
France
+33 (0)4 7382 5000
decovision@omerin.com
www.decovision-omerin.com

Decorated cable

Composition
Electric cables

Properties
Decorative and resistant to repeated use

Applications
Decoration, architecture, product design, etc.

Technical specifications
· Decovision standard reference for electric cables is H05VV-F 3 G, 0.75 mm^2, 300/500 V, HAR approved, compliant with standards HD 21.5, NF C 32-201-5, VDE 0281-5, NF C 32-081
· An UL-approved range of cables is also available
· There are 16 sections and 29 motifs in the standard range. Other types and/or sections can be made to order

The long-awaited opportunity to say goodbye to those ugly cables running around our houses is now here. This company has succeeded in perfecting a system for making electric cables part of the decor. The motifs resist the wear and tear of everyday life. This decoration technique can also be applied to flexible shower hoses.

1

2

1 & 2 The 'Techno' family consists of shiny cable patterns with glitters.

3 With 5 families of decorated cable patterns, glossy, gold and chrome looking cables belong to the 'metal' family.

48

3

Contact
Dow Corning Corporation
Corporate Center
PO Box 994
Midland, MI, 48686-0994
United States
deflexion@dowcorning.com
www.dowcorning.com/deflexion

Composition
Silicone and 3-dimensional spacer textile

Properties
Comfortable, flexible, protects against impacts, resistant to water and extreme temperatures, washable and breathable

Applications
High-performance apparel for sports such as motorcycling, snowboarding, skiing, and mountain biking, base-layer garments for sport and medical use, and personal protective equipment (PPE)

Technical specifications
· Available in rolls, easily cut for insertion into garments
· Can be used in several layers for particularly vulnerable areas of the body
· Can withstand 50 wash cycles at 40°C (104°F) and 25 at more than 90°C (194°F)
· Effective between -20°C (-4°F) and 40°C (104°F)

Deflexion™

Deflexion™ impact protection textiles are based on specially developed energy absorbing silicones. Deflexion™ S-Range technology is based on 3-dimensional textiles impregnated with silicones. These astonishing materials seem to behave differently according to the external environment. They can be very flexible and soft until they experience a violent impact. They then behave as if they were a hard material offering high levels of protection. Deflexion™ S-Range impact textiles, allow a motorcyclist or a skier, for example, to wear suitable clothing and move with no constraint or sensation of being armoured while being well-protected in the event of a fall. The energy absorbing properties of silicones can also be accessed via the thermoplastic Deflexion™ TP-Range.

1 The flexibility and breathability of this protective textile allows it to be integrated into close-fitting sportswear garments or gear worn close to the skin.

2 Detail of the Deflexion™ material.

3 Deflexion™ impact protection textiles could be considered for incorporation into footwear for protection of the sole, the heel or the top of the foot. This protection is imperative for athletes whose sport involves impact to the feet.

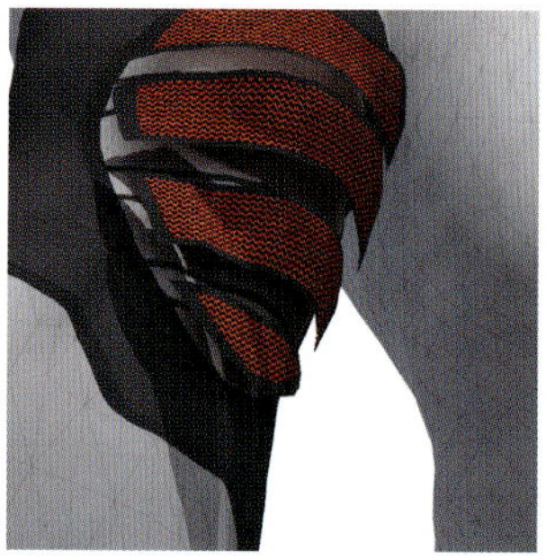

1

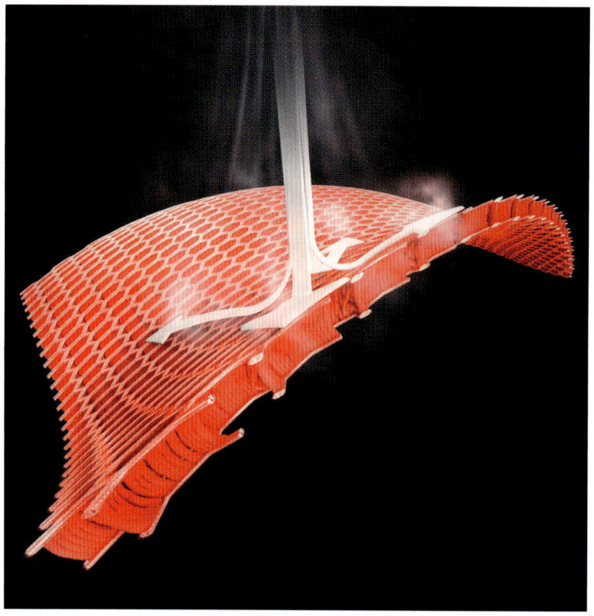

2

3

50

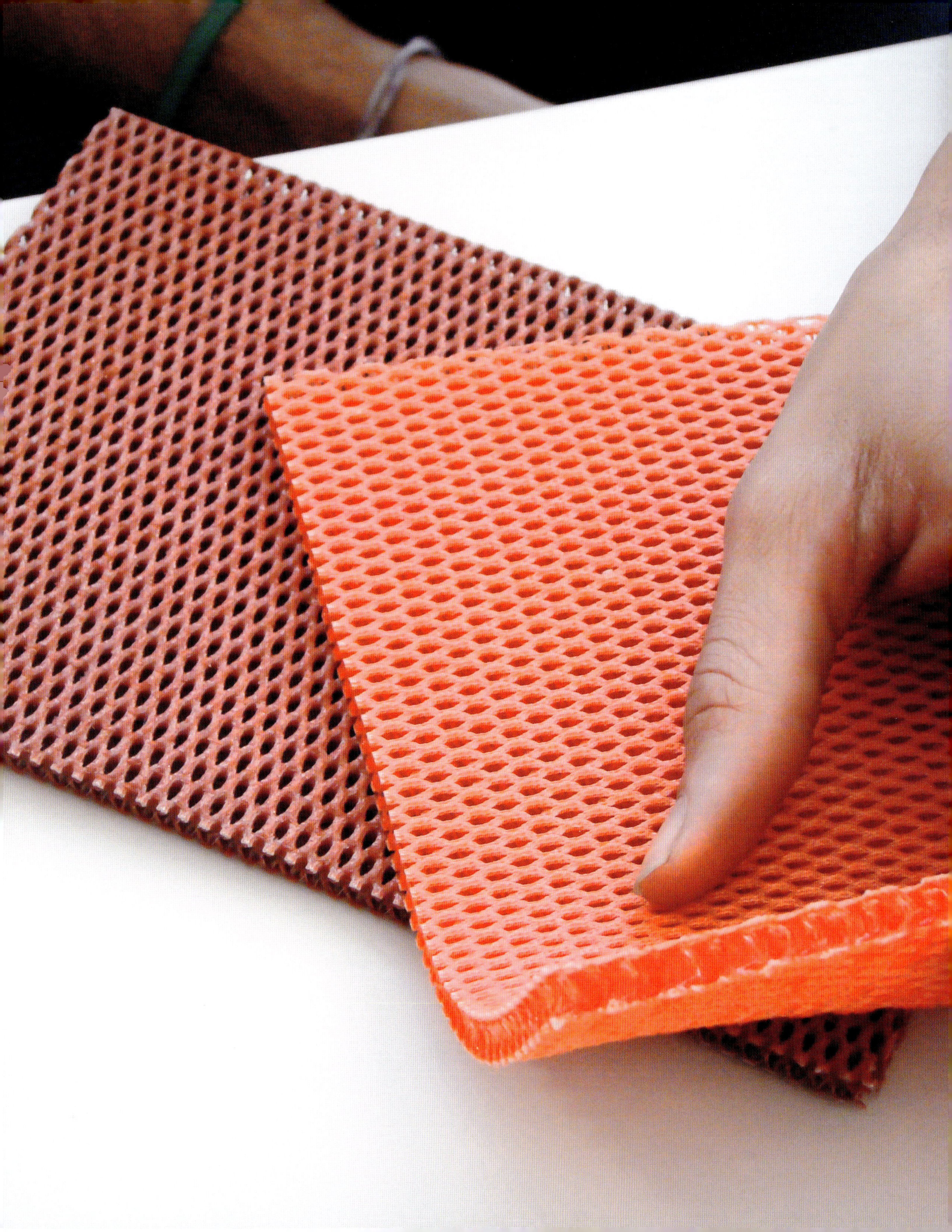

Contact

Coatings By Sandberg
856 N. Commerce Street
Orange, CA, 92867
United States
+1 (0)714 538 0888
www.cbs-dichroic.com

Dichro-patterns

52

Composition

Glass, with surface treatment based on metal oxides and quartz crystals

Properties

Decorative, very durable, proof against weather, water, corrosion, etc.

Applications

Originally for aeronautical use, now also found in architecture, tiles, glass bricks, swimming pools, sculpture, light fittings, but also in optical instruments, lasers, optical fibres, spectacles, etc.

Technical specifications

· More than 50 colours of coatings are available to be applied to various glass substrates. The possibilities are infinite
· Coating thickness: $< 1\ \mu m$

Vacuum deposition is used to 'grow' this ultrafine multi-layered coating based on metal oxides and quartz crystals as a crystalline structure on the surface of the glass. The number of colour layers can be as many as 30. The princi-pal of dichroism is simple: the glass transmits one colour and reflects another. The colours also change according to the viewing angle. The resulting colouring effect is therefore absolutely amazing.
Glass materials given this surface treatment can then be reworked hot without any colour distortion.

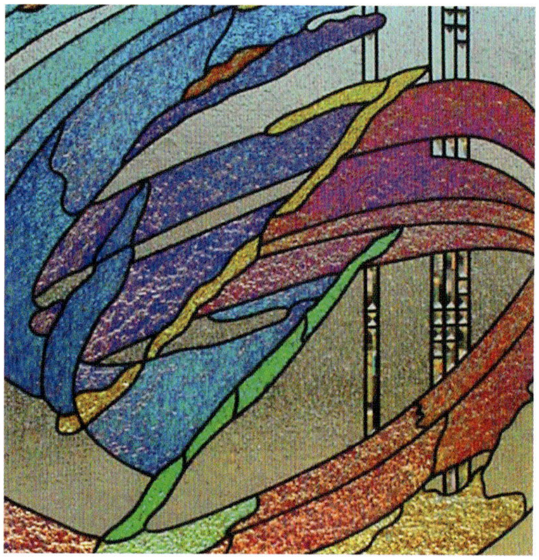

7

transcending the simple material aspects. It is essential to put the horse before the cart, because the environmental impact of a building or any object will be directly related to the initial approaches taken and, in a more anecdotal sense, to the choice of materials. Then again, it is obvious that it nevertheless seems necessary to recall that each project is individual and, in its own right, unique. In fact the answer as far as materials are concerned is part of a comprehensive analysis of the project's framework or its context, the way it is implemented, how it makes its mark in place and time, its purpose. No recipe, no 'one size fits all', but tailor-made because that's the way it has to be. A material may be unsuitable for one project but be the best possible compromise for another. Again, it is unrealistic to ignore a reality that is, above all, complex.

Eco-friendly materials must overcome the fact that they are something of a caricature in this debate because they are primarily food for thought. Even if certain criteria are emerging as vital (notions of intrinsic energy or renewable resources, for example), it is urgent today, and no doubt salutary, to re-consider our position and fight prejudice about materials that in themselves are also sustainable. No material is in itself intrinsically 'good' or 'bad' in absolute terms, as far as its ecological aspects are concerned. But on the other hand, while constantly having to make choices and decisions, we should make no economies in the use of one of the most natural, renewable and safe materials available: our own grey matter.

2

Contents

Material World 3

Innovative Materials
for Architecture and Design

Frame Publishers, Amsterdam
Directed by **matériO**

opportunity to change things profoundly. This crisis can have a cathartic effect if we know how to take advantage of this newfound freedom and the avenues which open up to enable us to 'get back to basics'.

In this context, more than ever, the existence of such a book is justified. In effect, it gives access to a wealth of suggestions and promotes a process for breaking down barriers and opening up. We thus hope to make a modest ripple in an ocean of inspirations, with its seething mass of innovations.

Ecology: prejudice about sustainables

Faced with real environmental imbalances caused by human activity, the pursuit of 'environmentally-friendly' material has in a very short time become the holy grail of every creative person and business: a requirement, seeming necessity, compulsory device but above all a fantasy. In effect, that description would clearly mean that there are intrinsically 'good' materials in existence (a notion generally attached to terms like 'natural', 'healthy', 'recyclable', 'made from renewable resources') whatever use is made of them and the other materials now doomed to be ostracised from society as lacking the qualifications to join the club, where all the members are 'very ecological materials'. This quest would be laudable if it were not simplistic, sterile and ultimately dangerous, leading to a stalemate for several reasons.

First of all, the reflex which makes us want to respond to current environmental issues primarily through the choice of materials must be resisted. This is going about the problem in the wrong way: materials must result from a comprehensive eco-design approach, which intrinsically greatly exceeds the practical aspects of materialisation. Think first about the nature of the project and how to respond to it, without thinking about what you already know and not hesitating to start with a clean slate free from basic premises. This allows a relevant, constructive solution to emerge, designer, artist) into a situation that is totally new in history. Now, more than ever, there needs to be perfect symbiosis between the leading edge of fashion, good taste and ethics. The choice is now so vast, the echoes sometimes so dissonant, that it's becoming a very complex matter to know where and how to manage projects. Even though this book does not in any way claim to solve all your daily challenges in selection of the right material for the right project, it remains a valuable tool to orientate yourself, to learn and to discover.

Times of crisis

The theory of evolution and, on a more human scale history, shows us to what extent the concepts of crisis, innovation and change are intimately linked and inseparable, feeding off each other. Innovation is stronger in times of crisis, destruction initiates creation, after death there is life - but a different life. In their infinite and legendary wisdom, the Chinese transcript for the word 'crisis' as an ideogram is based on the notion not only of threat but also opportunity.

Today's crisis seems largely beyond the simple field of finance and even the actual economy. We have a real crisis in society, calling into question the worn-out model that is exhausting our world, coupled with a desire to re-examine from every angle the very structures of our societies, their mode of operation, modes of production, consumption, the machinery allowing the whole to function, the distribution of wealth, the ends pursued.

We can suffer this crisis, put our umbrellas up and endure being buffeted, putting away our plans for the future while we wait for a respite. Or we can take advantage of this moment of wavering certainties to take up our freedom, no longer accepting ready-made ideas, throwing out conventional patterns, clichés, illusory tricks and trendy gurus and then becoming agents of this transformation, regaining from it our free will and our curiosity. There is now a perfect

4

Introduction

Third opus of its genre, *Material World 3* gathers together more than a hundred or so references for materials, semi-finished products or technologies selected by matériO. A careful choice covering a wide variety of characteristics and aesthetic qualities: modular, undulating wall coverings in mixed terracotta's, staghorn panels, fish skin leathers, ultra-thin batteries, crushed almond shell, LED lighting systems or photovoltaic textile. Spread around the whole world, their manufacturers are all different but offer completely new products, ready to be taken up by you the designer.

Every day, matériO identifies new products whose technical characteristics and/or aesthetic appeal are unusual and is always on the alert to technological advances, guaranteeing access for its members to a huge database with thousands of possibilities. *Material World 3* shows only a small part of it but whether it is for immediate use in a project you are working on, to add to an already bulging address book or to simply leave for future inspiration, the database contains many possibilities.

Increasingly attached to teaching the subject to some extent, and persuaded that innovation results largely from maximising the use of materials and technologies by having in-depth knowledge of them, matériO, the library of emerging materials, has in this book extracted fifteen basic concepts to help novices gain a better understanding of this complex world and remind the well-informed amateur of some basic elements that he or she never took time to verify. What do FSC, VOC, Cradle to Cradle™ and oxo-degradable mean exactly? What are damask steel, float glass or electroluminescence? What is the story behind the invention of Nylon, Velcro®, etc.? So many small texts that are available to the reader as informed reflections.

Our world evolves fast, materials and semi-finished products quickly appear and disappear. At the time these pages are being read, it is even possible that some of the materials referred to may no longer be available. The accelerating time we are living in plunges the creator (architect,

3

217

Biomimetics

Biomimetics is a methodology which seeks to artificially reproduce principals present in nature. Inspiration to be drawn from existing biological systems follows on naturally from this and an enhanced knowledge of natural phenomena and functions opens up the route to innovation for us. In particular, Janine Benyus, an American scientist, has insisted on this way of comprehending nature. Her book, *Biomimicry: Innovation Inspired by Nature*, published in 1990 is still a work of reference in this field. The biologist emphasises the following points:

- Nature relies on one main source of energy: solar energy
- Nature uses only the quantity of energy required
- Nature adapts form to function
- Nature recycles everything
- Nature bets on diversity
- Nature works from local assessments
- Nature does not 'go too far', limiting its own excesses
- Nature uses constraints as a source of creativity.

Nature becomes a model and a measuring instrument in the service of a better world management system. A good reason for preserving biodiversity. Examples of biomimetic technologies become ever more numerous: Velcro®, invented by Georges de Mestral, resulted from the close attention he paid to burdock balls; the lotus effect which reproduces the (microscopically) bristly texture of lotus leaves to recreate their water-repellent and self-cleaning properties on the surface of textiles; the blades of windmills which have crenellations, following the example of the fins of certain whales, to create swirl effects for greater efficiency or the giant termites which inspire architects to design buildings which are thermally regulated without additional energy input.

See notions: **Lotus effect** p.225 and **Velcro®** p.230
See catalogue: **C-Change™** p.30, **Mincor® PES** p.130, **Ottimat™** p.144, **Pure Bond® Hardwood Plywood** p.162 and **Stolotusan** p.184

218

Cradle to cradle™

The Cradle to Cradle™, C2C, certification mark arose from the vision of two men, American architect William McDonough and German chemist Michael Braungart. Their book, *Cradle to Cradle – Re-Making the Way We Make Things* – which appeared in 2002, defended hitherto unpublished theories in the field of ecology, not based on a reduction of consumption but rather on a revisiting of industrial processes. A biomimetic model where everything must be re-used, returned to the soil as biological, non-toxic nutriment or recycled 'technical' nutriment ad infinitum. No waste, one of the watchwords.

This holistic approach applies to every type of system (social, economic, architectural, etc.). It relies on a vision of human activities as combined metabolisms, potentially healthy and vigorous if they are fed by a controlled flow of materials, organic and synthetic nutriments. In this context, certain substances are to be banned once and for all, as in C2C list X, which groups teratogenic, mutagenic or carcinogenic elements for example. In contrast, list P welcomes as 'positive' non-dangerous substances and actively encourages their use. In the face of a decrease, the two leading lights of the movement defend intelligent consumption, based on continuous re-use of materials and objects. Products must be considered upstream of the creation process in order to guarantee several lives for them.

Numerous manufacturers have already taken the plunge to obtain this certification. They integrate notions for recycling their product into its production: a carpet manufacturer which recovers worn carpet tiles to reprocess them, a manufacturer of biodegradable cosmetics, compostable garments, chairs which can be dismantled to be recycled, etc.. Until whole towns in the Netherlands or in Germany are well on the way to becoming 100% Cradle to Cradle™ by means of an impressive zero waste policy.

See catalogue: **Greensulate™** p.86

219

Damask steel

Damask steel is legendary, long-considered indestructible. There were already traces in the Caucasus around 2500 years BC. At the time of the Crusades, damask steel swords were being made from steel with a high carbon content (between 1.5 and 2%) raised to 1200°C, with slow cooling to encourage the formation of lamellar patterns of an iron compound called cementite. The name damask certainly comes from this era of great voyages, in reference to the moiré-patterned damask material.

These days, damask steel is obtained by alternation of layers of soft and hard steel which is welded, drawn, heated, folded, twisted, welded, etc.. After grinding and polishing, patterns are revealed by applying acid to the surface of the metal. The bars obtained are sometimes slit with a chisel to obtain symmetric patterns before being assembled to form a blade, for example. The mechanical properties of this type of steel are comparable to homogeneous steels with equivalent carbon content. Damask steelmaking skill is not widespread, only a few master blacksmiths still practise the art. The pieces made are often unique pieces, in general artistic cutlery of high value. The blades are attached to handles of precious materials: mammoth ivory, precious woods or other rare materials.

Today, fabrication techniques are very highly developed, especially in countries like Japan where the long history of damask steel has never been interrupted. The greatest masters know how to create the most complex patterns, even script, by alternating layers and repeated working of the material. It is also possible to acquire 'blanks' of damask steel ready to be machined, for example for applications in jewellery or for small luxury inserts in more industrial objects.

220

Electroluminescence

There are different types of light emission, caused by a rise in temperature (incandescence) or by the action of chemical substances (chemiluminescence), for example. Electroluminescence (EL) is an opto-electrical light-generating phenomenon which was first observed at the beginning of the 20th century. In response to an electric current or a strong electric field, certain materials emit light. Among these are phosphorus, zinc sulphide doped with copper or silver, 'blue' diamond (diamond doped with boron), type III-V semiconductors, such as indium phosphide (InP), gallium arsenide (GaAs) and gallium nitride (GaN). There are also organic semiconductors. The principle of electroluminescence lies in the emission of photons by electrically-excited electrons, during recombination of these electrons with holes in the material (generally a semiconductor).

Phosphorus-based electroluminescent surfaces, very thin and flexible with low energy consumption, are frequently used as backlighting for liquid crystal displays or for information displays. Easily produced colours are green, blue, turquoise, yellow, orange and white. Pigments emitting intense pure white are still at the experimental stage. In the meantime, EL sheets giving out white light have a minimum of two pigments which add together to give white. When de-energised however, these light sources are pink in colour. This phenomenon of light emission by a semiconductor is the basis of light emitting diode (LED) operation.

See catalogue: **ReLight** p.170

221

Fibre-reinforced concretes

Fibre-reinforced high-performance concretes are the latest in a generation of concretes with exceptional performance. Progress in this field relies on increased proficiency in the domain of additives, methods of formulation and use of ultrafine additives (silica fume). Construction techniques are also going through a real revolution. Today, it is possible to envisage the elimination of passive reinforcing material in structural elements thanks to the tensile and compressive strength, ductility and consequent durability of fibre-reinforced high-performance concretes.

These specific concretes are cement-based materials with superplasticising additives, reinforced with metal, polymer or mineral fibre (at high rates in the order of 2 to 3% of fibre by volume). The fibre generally has a diameter from 0.1 to 0.3 mm and a length between 10 and 20 mm. Metal fibre is used more for works which require high mechanical strength, while polymer and mineral fibre are used where the aesthetic aspect is important. The formulation of these new concretes seeks to reduce the porosity of the material (which means less shrinkage for example) and to optimise its granular network (for a more compact, stronger mixture). As a function of the fibres and their size these materials can be, among other things, very resistant to fire and impact. The concretes are to be found in numerous domains, from civil engineering (beams, pipes, offshore constructions, silos, slabs, roofing over large spans, etc.) to building (beams, large flooring spans, etc.) as well as furniture. In general, on construction sites there need be no vibration and reinforcing, improving things for building workers.

The properties of these concretes enables us to imagine very slender constructions, perhaps even open structures like lacework or latticework. Their fine microstructure makes them especially suitable for moulding and surface finishes straight out of the mould can be very clean. The 'skin' of these concretes can therefore be very smooth, silky, mat or shiny, but can also easily take impressions.

See catalogue: **Arty** p.18, **B-ton Scintillant®** p.22, **Concrete Pillow** p.36 and **Textured Concrete** p.190

222

Float glass

The manufacture of flat glass reached a major turning point in the 1950s when Alastair Pilkington perfected the so-called 'float' technique discovered by Henry Bessemer in the 19th century. Production, until then a craft, could then develop fully on a large scale, guaranteeing a product with exceptional optical qualities and cheaper than previously. The float principle is as follows: a ribbon of molten glass at approximately 1,000°C – composed of sharp sand, lime, soda and various oxides – flows over a bath of liquid tin. The very smooth surface of the molten metal imparts a perfectly flat surface to the glass. The ribbon of glass is then cooled slowly until it is completely hardened, before being annealed. The resulting product has constant thickness and perfectly polished surfaces. The ribbon is then cut into plates of the required length.

The thickness of the float glass available today varies from 0.4 to 25 mm and it is possible to add colour to the initial mixture, grey, green, bronze or blue for example, to obtain self-coloured float glass. However, the majority of coloured glasses that we encounter are laminated, with a central core enclosing a coloured film.

80% of flat glass produced in the world today is float glass. The major sectors for use are building, architecture and automobiles. Flat glass continues to be produced by drawing, rolling or pouring into moulds. Glass produced by these methods has irregularities which give it a certain charm, appreciated in many situations.

See catalogue: **Glasswood** p.78 and **Semi-mirror Glass** p.174

223

FSC

The Forest Stewardship Council (FSC) was created in 1993 after the Rio Earth Summit of 1992. This international non-governmental independent and non-profit-making organisation consists, in equal sections, of the economic players in the wood industry (forestry industries, distribution groups, etc.), social players (unions, people's representatives, etc.) and environmental players (NGOs such as the WWF or Greenpeace).

The FSC's objective is to protect forests against often untimely deforestation, in particular in tropical zones, and in doing so to encourage forestry management that is socially, economically and ecologically responsible. The FSC provides an eco-label which guarantees that a wood or paper product really has been sourced from a certified forest. Certain species of wood are thus classified today by the FSC as being seriously endangered. Use of the following should therefore be avoided: mahogany, merbau (also known as kwila), sapelli, sipo or teak. The following are currently considered safe for exploitation: oak, spruce, maple, ash, beech, pine or wild cherry.

An eco-label is obtained via an inspection by an independent certification body, which verifies that the points defined above have been properly observed. In effect, the FSC drew up a list of 10 principles and 56 criteria valid for all types of forest. These serve as a reference for drafting a forest management standard: transgenic species are banned, pesticides are prohibited or very strongly regulated, animal and vegetable biodiversity must be preserved and the rights of forest peoples must be respected, for example. Two types of certification can be issued: certification of forestry management and certification of the chain of inspection.

Finally, it should be noted that this eco-certification has certain limits: if there is a worldwide requirement for additional paper, eco-certification will not be able to affect the decision of certain countries to convert their fields into forests, for example. Which could have catastrophic repercussions on environmental, social and economic stability.

See catalogue: **Pure Bond® Hardwood Plywood** p.162

224

Liquid crystals

For a long time man considered matter through the prism of three time-honoured states: solid, liquid, gaseous. It was finally only recently in the 19th century that he started to take a close interest in intermediate states.

Liquid crystals give a perfect illustration of a badly defined state of matter: they combine the properties of a conventional liquid with those of a crystalline solid. This particular state is also described as a 'mesophase' state. In polarised light under a microscope, a liquid crystal reveals distinct zones of different textures in which the molecules are oriented in different directions.

Two major types of liquid crystal are distinguished: thermotropes in which a phase change occurs as a function of temperature and lyotropes which are sensitive to temperature and to the concentration of molecules in a solvent. It was the study of cholesterol by Friedrich Reinitzer in 1888 which would mark the start of the study of mesophases. He was able to observe how the optical opacity of cholesterol esters and their iridescent colours change when their temperature varies, taking them from the crystalline phase to the liquid isotrope phase. In 1968, George H. Heilmeier, an American researcher, perfected the first liquid crystal display (LCD). This was commercialised five years later when a stabilised system had been developed by Seiko for a revolutionary consumer watch.

Today, there are many applications of liquid crystals, from the famous LCD screen to forehead digital display thermometers with windows which become opaque or a welder's helmet which darkens in less than 1ms when it detects a high light level.

See catalogue: **SmartGlass™** p.176

225

Lotus effect

The image of a lotus leaf with a perfect drop of water nestling in a hollow is emblematic of this plant. Symbol of purity in the Buddhist religion, the self-cleaning properties of the lotus have been known for more than 2000 years. If we imagine intuitively that the ever immaculate surface of the leaf must be extremely smooth, the reality is quite different. Meticulous observation of the texture of a lotus leaf reveals that its microscopic roughness is the origin of its super-hydrophobic properties. An electron microscope photograph reveals that the surface of the leaf has a micro-texture, on a scale between 1 and 10 μm and consisting of hydrophobic waxes. Water droplets are not able to adhere to the surface because they have very little contact with it. Essentially, they remain in the air as they do in a cloud and roll over the leaf, carrying away any dust present en route. There are therefore two effects, waterproofing and self-cleaning.

This surface state, microscopic or nanoscopic, was demonstrated in the 1970s by botanist Wilhelm Barthlott, who in fact has received numerous prizes for his research. Several plants show this effect, for example nasturtium or cabbage leaves, but also the carapace of certain insects.

In the 1990s the first biomimetic applications of the phenomenon appeared: completely waterproof fabrics, water-repellent concretes and paints or self-cleaning windows. However the effect is not permanent and the surfaces will possibly have to be re-treated to maintain their interesting properties. It remains to be added that many so-called 'easy to clean' products come onto the market but there are few for which the description is appropriate or have any connection with the lotus effect.

See notions: **Biomimetics** p.217
See catalogue: **Mincor® PES** p.130 and **Stolotusan** p.184

226

Nanoparticles

We have all heard of nanotechnologies. As a function of the scale at which material is observed and manipulated, a change in its properties occurs and especially so as the nanometric scale approaches. This is explained in part by the fact that the smaller it gets, the more important the role played by the material surface, while at the macroscopic scale the small thickness of some surface atoms is negligible. A nano particle is a group of atoms, from a few hundred to a few thousand, with a size between one and 100 nm. This is therefore not the smallest indivisible particle in existence, but its small scale involves changes in its physical and chemical properties that are worthy of interest. Nano materials are classified according to their size in three-dimensions: Fullerenes and particles are entirely within the nanometric domain, nano tubes have only two nanometric dimensions, while thin films (planes) have only one nanometric dimension, their thickness. Industrially, nano particles are obtained either by agglomeration of atoms (bottom-up method) or by degradation of material (chemical, physical-chemical, biological or mechanical methods).

The increasing spread and commercialisation of nano particles raises crucial questions of health and safety. Their small size makes them particularly capable of penetrating living organisms and we currently have very little knowledge on the effects of such widespread distribution. There are standards emerging in some countries but legislation is not yet clear on the subject.

Several types of nano materials are in use today: carbon nanofibres in the field of electrical conductivity, for artificial muscles or for hydrogen storage, nano crystals such as diamond in the manufacture of new microprocessors , nano pigments, ceramic nano powders for surface treatment of other materials, etc..

See catalogue: **Mincor® PES** p.130 and **TiColor** p.196

227

Nylon

Nylon designates a particular type of polyamide: polyamide 6-6 or polyhexamethylene adipamide. This polymer, invented in 1935 by Wallace Carothers, an employee of the American DuPont company, is obtained by hot condensation of molecules. As a thermoplastic, it can therefore be injected or extruded.

The invention came in the turbulent period leading up to the Second World War and uses for the new material were no doubt sought to replace other materials that were difficult to obtain in the war, such as silk for example. The origin of the name is hazy, between the name of the inventor's wife and a corruption of the word 'Norun' (i.e. does not develop ladders), changed into 'Nuron' which became Nylon, for example.

Today, the word Nylon is used to designate any type of polyamide even if in reality it refers to a particular member of this family. Nylon fibres have almost unequalled resistance to friction and their applications underline this: toothbrush bristles, parachute material (rip-stop Nylon) or the well-known Nylon hosiery. Nylon is also found, in solid form, in the manufacture of small mechanical parts such as toothed wheels in the automobile or food industry sectors. In a matrix of composite material, reinforced with glass or carbon fibre, Nylon becomes even stronger and its heat resistance (melting point at 256°C for Nylon fibre) allows it to compete with metal in proximity to automobile engines. Its low absorbance also allows it to be used, in fibre form, in numerous quick-drying garments: sportswear, raincoats, swimwear, etc.

See catalogue: **Annemette Beck** p.12, **Cork Fabrics** p.38 and **Freek** p.72

228

Oxo-degradable

In the context of sustainable development in which we now find ourselves, materials are often associated with notions of being 'bio-degradable, compostable, bio-fragmentable, or oxo-degradable'. Today, these notions are defined precisely by European standards.

Bio-degradable: the bulk of the initial material must be 90% degradable, therefore any waste residues must represent a maximum of 10% of the initial material tested. The result of bio-degradation must not have ecotoxic effects.

Compostable: Waste residues must represent a maximum of 10% of the initial mass of the material tested, the size of the waste residue particles must be less than 2 mm (disintegration), the absence of negative effects on the composting process and the result of composting must not have ecotoxic effects on the compost.

Bio-fragmentable: Materials described as bio-fragmentable, fragmentable or oxo-fragmentable are in general mixtures of synthetic polymers (polyethylene type) with vegetable or mineral additives. The life end of these materials manifests itself by bio-degradation of additives and by a physical (visible) degradation without molecular disintegration of any synthetic elements. These materials do not meet the requirements set by the standard in force.

Oxo-degradable: The term 'oxo-biodegradable' is sometimes used. Generally, thermoplastics with additives (polyolefins containing an oxidising agent, such as dithiocarbamates of iron, nickel or manganese or nickel stearate to promote degradation). They are in fact not very bio-degradable, but rather they are 'oxo-degradable' with a reduction in molecular weights with fragmentation of the material. The long term effects of accumulation of polyolefins in the soil is not known (even if the material becomes invisible to the naked eye). In addition, the first additives used contained heavy metals. Dithiocarbamates are also toxic even if they do not contain heavy metals. These substances do not at present meet the standards or labels on the bio-degradability of materials.

See catalogue: **Bioxo™** p.26

229

PCM

The functioning of phase-change materials (PCM) is based on simple physical principles of change of state. In a fusion phase for example, above a certain temperature characteristic of each material, they liquefy by absorbing energy and release it in the form of heat when the temperature falls.

There are different types of PCMs: mineral compounds such as hydrated salts, organic compounds such as paraffins and fatty acids or eutectic compounds (mixtures of two organic or inorganic elements. A well-known mixture being water+salt during winter, to lower the freeze temperature on roads). These amazing materials hold the promise of thermal regulation for buildings. In effect, the phase change can take place between 19°C and 27°C, temperature limits set for winter and summer comfort, and spaces surrounded by PCMs can thus have the ambient temperature reduced or increased by 3 to 5°C. PCMs can thus take over from energy-devouring air-conditioning systems. Too warm, PCMs absorb excess heat and help to lower the temperature, too cool, they restore the energy previously absorbed in the form of heat, causing a small rise in temperature.

These phase-change materials are now not only present in paints, ceramics, flexible polymer coatings to be applied to the walls as an under-layer, but also in textiles to offer optimum thermal comfort in sports clothing for example.

See catalogue: **Energain®** p.66

230

Velcro®

Every time Georges de Mestral went for a walk with his dog, the animal came home covered in small burdock balls. He then needed a lot of time and patience to remove these little pieces of vegetable material, which were securely hooked into the animal's hair. How did they hold on so well? Examination under the microscope revealed that burdock balls have tiny curved hooks which catch very effectively on anything coming past: fabrics, hair, clothing. Furthermore, these hooks are so flexible that they deform when they are pulled up, regaining their shape and their ability to hook on again immediately. Georges de Mestral suddenly realised that by reproducing and imitating these hooks, he could devise a new closure principle.

Thus was Velcro® invented in 1948. Consisting of two strips, one covered with 'velvet' and the other with hooks, which stick together firmly and indefinitely when flattened against each other. The name is quite simply a French combination of 'Vel' for the velours (velvet) part and 'cro' for the crochet (hook) part. This fastening system has since had numerous imitations and has given rise to lots of variations. Today there are hydrophobic, nano-Velcro®, conducting Velcro® systems, etc.. This system is a good example of biomimetics where nature has pointed us to a simple way of structuring material for impressive fastening capability.

See notions: **Biomimetics** p.217
See catalogue: **Wall Carpet** p.204

231

VOC

There is no clear definition of Volatile Organic Compounds (VOC). These are organic substances consisting of carbon and at least one of the following elements: hydrogen, oxygen, sulphur, phosphorus, silicon, nitrogen or one of the halogens. They have a marked ability to vaporise at ambient temperature and pressure and escape into the atmosphere. Whether they have a natural origin (emission by vegetation) or industrial origin (solvents, refining, etc.), some of them are toxic. They are in effect 'greenhouse' gases. Among the best known are: butane, propane, ethanol (90% alcohol), acetone and paint or ink solvents.

The blacklist varies according to the country. In Canada for example, methyl chloride is excluded from the list of VOCs while it is included in France. Today, the emission of VOCs is regulated, but not in the same way everywhere. In the European Union, a VOC is an organic compound whose boiling point is not higher than 250°C at 101.3 kPa (1.013 bar) pressure and which can cause harm. All techniques which necessitate manipulation and production of hydrocarbons cause the emission of VOCs. Road transport for example plays a major part in the emission of these compounds.

Currently, there are some reductions to be noted, thanks among other things to the Kyoto protocol of 1997 and the efforts of the automobile industry. But VOCs are also present in our homes, given off by paints, adhesives and products used to treat wood, for example. Certain plants are known for their ability to absorb pollutants.

See catalogue: **Ecocarat Takehigo** p.60 and **Placo Impact** p.154

Colophon

Material World 3
Innovative Materials
for Architecture and Design

Publisher
Frame Publishers

Directed by
matériO (www.materio.com)

Author
Elodie Ternaux

Production
Marlous van Rossum-Willems

Graphic design
Atelier MAJi

Translations
TransL Vertaalbureau
(Clive Pygott and Liz van Gerrevink)

Photographer
Maroussia Jannelle
(material photo's)

Prepress
Edward de Nijs

Printing
D'Print

Trade distribution USA and Canada
Consortium Book Sales & Distribution, LLC.
34 Thirteenth Avenue NE, Suite 101
Minneapolis, MN 55413-1007
T +1 612 746 2600
T +1 800 283 3572 (orders)
F +1 612 746 2606

Trade distribution Europe and South America
(excluding the Netherlands, Belgium, France and Turkey)
Birkhäuser GmbH, Basel
P.O. Box, 4002 Basel, Switzerland
www.birkhauser.com
sales@birkhauser.ch
ISBN: 978-3-0346-0754-4 (Birkhäuser)

Distribution rest of world
Frame Publishers
Laan der Hesperiden 68
1076 DX Amsterdam
the Netherlands
www.framemag.com
distribution@framemag.com

The Koninklijke Bibliotheek lists this publication in the Nederlandse Bibliografie: detailed bibliographic information is available on the internet at http://picarta.pica.nl

Printed on acid-free paper produced from chlorine-free pulp. TCF ∞
Printed in China

987654321

ISBN: 978-90-77174-26-5

Contact
Robatex
Prattelsackstrasse 25
52222 Stolberg
Germany
+49 (0)240 2124 6694
info@robatex.de
www.lucem.com

Dot99®

Composition
Concrete with optical fibres

Properties
Decorative, luminous and strong

Applications
Interior architecture and signage

Technical specifications
· Dimensions: 500 x 1000 mm up to 1000 x 1500 mm. Dimensions up to 2500 x 1000 mm on request
· Density approximately 2300 kg/m³
· Colours: white or grey, and other colours possible on request
· All surfaces are polished to a shiny finish and supplied impregnated with a water repellent

You will certainly recognise a translucent, multi-layered concrete, composed of concrete and optical fibres. The same formula for this material, but here the optical fibres are arranged in the concrete so as to create a motif or a luminous message. The concrete then becomes a potential communications medium. Creations can be made to measure: the client places his order via a data file, which is then 'woven' into the concrete. A light source is placed behind the device to energise the light display thus created.

54

1

2

1 Lucem cladding on a counter.

2 Lucem cladding used as a partitioning wall/blinds.

Contact

Duralmond
Avda. Región Murciana
Parcela ITC-2/3
Pol. Ind. Base 2000
30564 Lorquí Murcia
Spain
duralmond@duralmond.com
www.duralmond.com

Duralmond

1

This material consists of a synthetic resin loaded with crushed almond shells. Almond shells contain lignin and cellulose and as a result the mixture has performance and properties mid-way between plastics materials and wood.

Composition

Almond shells and synthetic resin

Properties

Use of renewable materials, until now not used to advantage, worked like a thermoplastic, with numerous colours and surface finishes possible

Applications

Bathroom furniture and wall and ceiling coverings

Technical specifications

· Impermeable to water
· High resistance to UV
· Resistant to insects
· Good acoustic and thermal properties
· Fire resistant
· Variable (controllable) density: from 175 to 700 g/l

2

1 & 4 A cylindrical section, 30cm in diameter, with the cut section slightly concaved. The Pastilla gives every wall an extraordinary look and feel.

2 The Orange Tree's courtyard of the EME Cathedral Hotel in Sevilla is surrounded by the impressive latticework of Duralmold.

3 The rectangular tile Cantos Rodados, from the traditional decoration product group, is imitating pebbles on a bed of plaster.

56

3

4

Contact
Coverings Etc
7610 NE 4th Court
Miami, FL, 33138
United States
+1 800 720 7814
info@coveringsetc.com
www.coveringsetc.com

Eco-Cem®

Composition
Concrete and wood pulp

Properties
80% Portland cement (calcium silicate and silicon dioxide) and 20% cellulose fibre (wood pulp)

Applications
Floor and wall coverings

Technical specifications
- Dimensions: 3048 x 1220 mm x 6, 13 or 38 mm (slabs) / 1220 x 610 x 13 mm and 1220 x 1220 x 13 mm (tiles) / Other sizes cut to order
- Eco-Cem® materials are stable
- Reaction to fire: 0
- Colours: Celadon Green, Clove Brown, Coal Grey, Deco Coral, Natural Grey, Safari Brown, Tibet Gold and Windsor Grey

These concrete panels contain wood pulp, using wood industry waste to create a contemporary material of industrial design standard which is warm to the touch. This company is on an environmentally-friendly route, in particular with the use of recycled materials. Their product ranges (glass, stone, concrete, etc.), all under the heading of ECOverings®, offer high-performance composite materials, based on material that is natural, recycled, recyclable, with conservation of natural resources in mind.

1

2

1 At the Boffi showroom in Washington DC the Eco-Cem® contemporary cementitious boards are used on the wall and floor.

2 An example of the Eco-Cem® Miami showroom.

58

Contact

INAX Corporation (HQ)
5-1 Koie-Honmachi
479-8585 Tokoname-Shi, Aichi
Japan
+81 (0)569 36 0360
www.inax-ecocarat.com

60

Ecocarat Takehigo

Composition
Ceramic

Properties
Controls humidity, deodorises and absorbs toxins

Applications
Building, individual houses, etc.

Technical specifications
- Can absorb a maximum of 9 g/m²
- Capable of absorbing/releasing moisture 4 to 5 times better than diatomaceous wall surfaces, and 15 times better than hygroscopic wall paper (based on tests conducted by INAX)
- Ecocarat contains no harmful substances, and generates no formaldehyde
- The material is available in a variety of rectangular shapes and various colours: imitation brick or stone, sand effects, etc.

Its micro-perforations absorb possible excess humidity and release it again if the atmosphere becomes too dry. In addition, it absorbs bad odours and some toxins such as formaldehyde, toluene and other VOCs (Volatile Organic Compounds). Humidity regulation also inhibits the development of mites and moulds.

See notions: **VOC** p.231

1

1 Ecocarat softens the effect of light, making it perfect for bedrooms where relaxation is paramount. It also helps eliminate odours, ideally suited for master bedrooms with en suite bathrooms.

2 & 3 Highlight your living room or bedroom in a beautiful mosaic to foster a relaxing, refreshing atmosphere.

2

3

Contact

Gebrüder Jaeger
Otto-Hahn-Strasse 7
42369 Wuppertal
Germany
+49 (0)202 24656-0
sales@jaeger-ttc.de
www.jaeger-ttc.de

Elast Hydro 26640

Properties

Elastic and water-repellent

Applications

Fashion, accessories, footwear, architecture, events, etc.

Specialised in the supply of elements for footwear and leather goods, this company offers, among other things, completely water-repellent tape, straps and other elastic products. Water forms beads and slides over the material without being able to penetrate the fibre.

© greg - Fotolia.com

Contact

Poesia
Via Boscalto 13
31023 Resana (TV)
Italy
+39 04 2343 1056
info@mattonepoesia.com
www.mattonepoesia.com

Elle

64

Elle is a glass brick which can have various shapes, colours and sizes. By combining different bricks, as in a well-known assembly game, partitions with astonishing geometries can be obtained.

Composition
Glass

Properties
Decorative, luminous, coloured and strong

Applications
Partitions, walls, etc.

Technical specifications
- Dimensions of bricks: 247 x 117 x 53mm
- Bricks have compressive strength of 2,500 kg/cm^2
- Scratch-proof (Mohs 7 certification)
- Resistant to wind, snow, frost and rain if correctly installed
- Resistant to mould, dust layers and chemical products (except acid)
- Low maintenance (just a wet cloth)
- Reusable
- 100% recyclable
- Non-toxic, does not absorb odours or liquids
- Fire-resistant (REI 60 certification)

Contact
Dupont de Nemours
+1 302 774 1000
www.dupont.com

Energain®

Composition
Phase-change materials

Properties
Thermal regulation, fine temperature control and energy-saving

Applications
Walls, floors and ceilings

Technical specifications
· Thickness: 5.26 mm
· Width: 1000 mm
· Length: 1200 mm
· Weight: 4.5 kg/m^2
· Aluminium thickness (sheet): 130 μm
· Aluminium thickness (edges): 75 μm

This energy-saving system is based on the use of phase-change materials (PCM). A panel is 60% paraffin wax, 40% copolymer. The paraffin wax is encapsulated in the copolymer by a specific technique. In alternating phases of melting and crystallisation, storing or releasing energy, the material will reduce temperature variations, increasing comfort.

See notions: **PCM** p.229

Contact
Moderne Methode
5 rue de la Bizière
77165 Saint Soupplets
France
+33 (0)1 6436 1526
info@moderne-methode.com
www.moderne-methode.com

Extreme Concrete

Composition
Concrete

Properties
Pliable while setting, fast out of the mould, soon ready for polishing and strong

Applications
Furniture (counters, kitchen worktops, tables, etc.)

Technical specifications
Strength up to 72 MN/m^2

This concrete offers several completely new properties: it can be released from the mould in about four hours, polished after eight hours and, above all, is pliable: in effect, extreme concrete is flexible during the setting phase and can then take on unique shapes, modifying our conventional perceptions about this material.

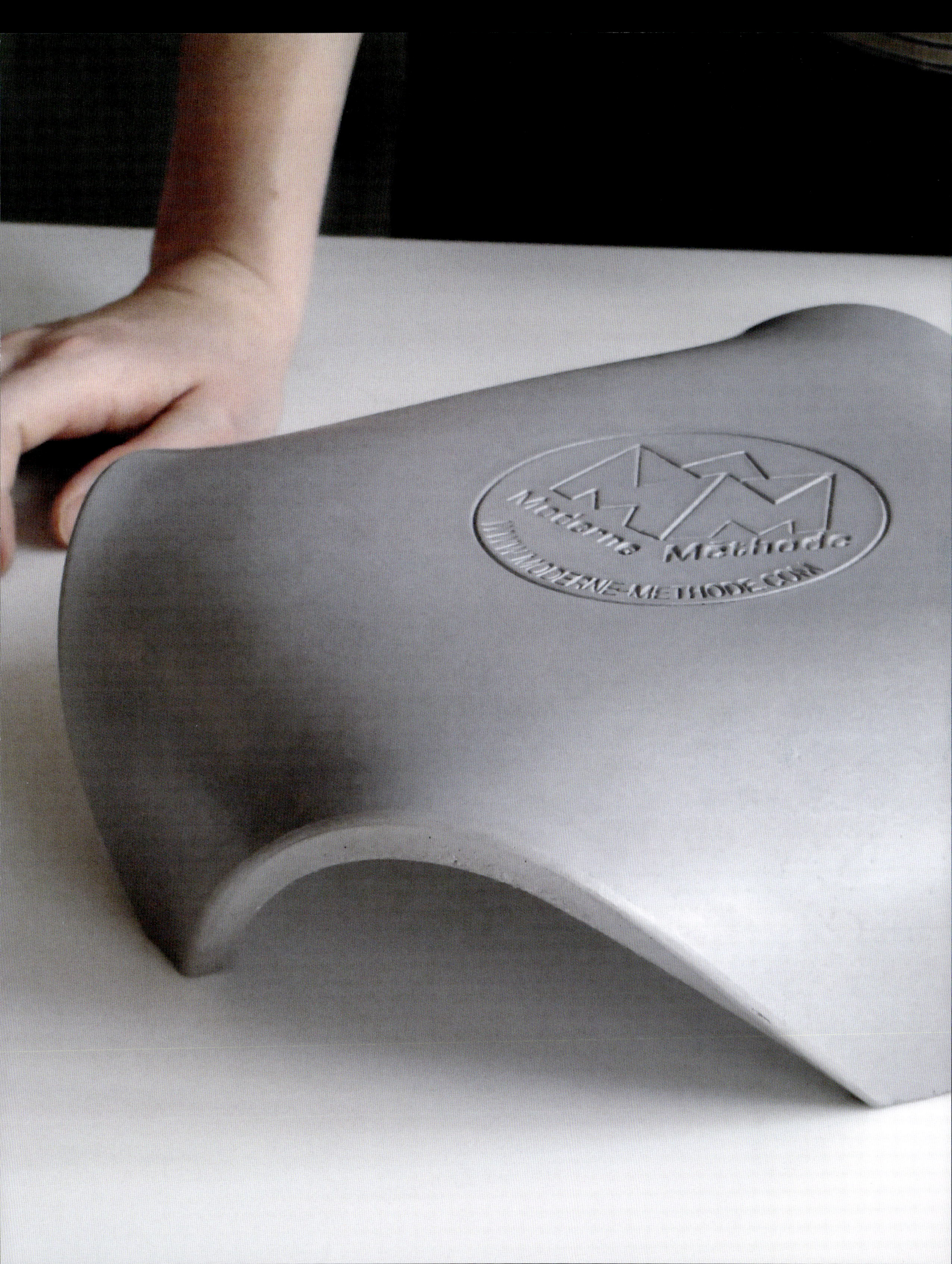
Moderne Méthode
WWW.MODERNE-METHODE.COM

Contact
EXYD™
Menzinger Strasse 113A
80997 Munich
Germany
+49 (0)89 8103 9475
info@exyd.com
www.exyd.com

EXYD™

Composition
Stainless steel, copper, aluminium, brass, etc.

Properties
Decorative, with possibilities for various forms, and durable

Applications
Façades, and external and internal coverings

Technical specifications
Dimensions according to requirements

These metal panels in stainless steel, aluminium or copper, formed in two or three dimensions as elements of façades or interior fittings are available as different series, some fully available, some still in development. The different product lines offer embossed reflective surfaces for light to play on 3D metal surfaces. The effect of light falling on these surfaces and the 'movement' created is quite astounding.

1

2

1 The six-metre high reception hall of the Waterfront Building in Stockholm, Sweden was fitted with an individually designed metal ceiling. EXYD™ developed the surface design for the ceiling based on the crackled effect on the surface of a half-frozen lake, and produced the panels.

© Michael Perlmutter

2 During the refurbishment of a residential complex in Dortmund, Germany the architects selected Product Line M to wrap part of the building's façade in a metal envelope.

Contact

Freek
Nylonstraat 7
8281 JX Genemuiden
the Netherlands
+31 (0)38 337 2014
hello@freek.nl
www.freek.nl

72

Freek

Composition

Nylon and polyethylene fibres

Properties

Decorative, water-resistant, UV resistant and easy to clean

Applications

Terraces, balconies, gardens, patios, etc.

Technical specifications

Dimensions: 1500 x 2000 mm, 2000 x 2000 mm, 2000 x 3000 mm, 2000 x 4000 mm, and other dimensions on request

Freek is a rather unusual fitted carpet, with acid colours, designed for outdoor use. It is manufactured from a combination of two types of fibre: Nylon, very strong, as a base layer, providing volume and longer polyethylene fibre giving a soft, smooth feel to the finished product. An integral underlay is a synthetic foam incorporating air pockets. The carpet is thus more comfortable and the surface underneath is all the better protected against scratches.

See notions: **Nylon** p.227

1

2

3

1 & 3 Freek carpets are water and UV resistant and therefore ideal for all-weather use outdoors.

2 The striped multi-coloured carpet shows a choice of available colours (15 different colours).

Contact
Nicolas Barondeau
58 rue de colombiers
34670 Baillargues
France
+33 (0)6 09 12 62 37
nicolasbarondeau@hotmail.fr

Fusiolam

Composition
Glass brick

Properties
Decorative, strong, transparent and stratified effect

Applications
Interior architecture, partition walls, windows, etc.

Technical specifications
Many possibilities of shape and length

These bricks are made from laminas of glass fused together, presenting a distinctive aesthetic effect when light plays on them. They are handmade and therefore can be manufactured in any dimensions. Their very specific look brings rhythm to architectural spaces.

74

Contact
Biothys (HQ)
Gewerbestrasse 6
77731 Willstätt-Eckhartsweier
Germany
+49 (0)78 5498 9540
contact@biothys.com
www.biothys.com

Gelactiv®

This company specialises in producing technical responses to olfactory nuisances by chemical and biochemical means. The Gelactiv® deodorising system effectively inhibits the main olfactory nuisances in all confined spaces and offers a sanitary guarantee. The product is in the form of sheets based on polymer loaded with the neutralising (non-masking) active principles, a mixture of natural and/or synthetic vegetable essences. This technique does not generate any nuisance: there is no liquid spray, no diffusion of dry vapours, no risk of dispersion of pathogenic germs, and no dangerous products.

Composition
Polymer, natural and synthetic vegetable essences and metal supporting lattice

Properties
Deodorising, easy maintenance, no electrical installation necessary and continuous targeted treatment

Applications
Local authorities, hotels, health providers, etc.

Technical specifications
· Available as three active compositions, each available in three sizes
· Each type of sheet, Gelactiv® SHK, Gelactiv® NHK and Gelactiv® BHK, is designed as a function of the gas to be dealt with, whether it relates to bad odours from urban effluents, industrial effluents or a mixture of these
· A metal lattice inserted into the moulding makes the sheet more rigid
· On average, the sheets must be renewed three or four times per year

Contact
Dalbergia
41 rue de la Fontaine au Roi
75011 Paris
France
+33 (0)1 5336 7633
contact@dalbergia.fr
www.dalbergia.fr

Glasswood

The laminated glass panel: a stylish veneer, with a finely-perforated motif, is inserted between two sheets of glass, light passing through the composite material then reveals a pointillist design. Numerous motifs can be produced, allowing these glass/wood panels to be individually customised.

See notions: **Float glass** p.222

Composition
Laminated panels composed of wood veneer enclosed in glass

Properties
Decorative, attractive light play and several motifs possible

Applications
Walls, screens/room dividers, light fittings, etc.

Technical specifications
· Maximum dimensions of glasswood panel: 3000 x 1300 mm
· Minimum thickness of glass: 3 mm (no maximum thickness). - Each lamina can have a different thickness
· Thickness of wood veneer: between 0.3 and 0.7 mm (more or less transparency)
· All types of glass can be used on one or both faces (glass which is annealed, toughened, extra-white, anti-reflective, water-repellent, etc. Note: 'extra-white' glass is a float glass without the normal hint of green)
· Can be made to withstand damp conditions

78

80

Glaszone

Contact
Glaszone
Werschenschlag 1
3531 Waldhausen
Austria
+43 (0)28 2688 110
info@glaszone.com
www.glaszone.com

Composition
Crystal and glass

Properties
Decorative, light play, sound and heat insulation and mechanically stable

Applications
Interior architecture (floors, walls, screens, etc.), furniture, signage, etc.

Technical specifications
· Colours: green, light blue, dark blue, yellow, red, clear transparent, amber, black and other colours on request

Splinters of lead-crystal glass are enclosed between two plates, also glass, to form a panel, on which there is an attractive play of light. The panels also provide sound insulation as a bonus. The splinters are mixed with a transparent binder stable in UV. The product thus has high mechanical and thermal stability. This technique allows creation of motifs 'suspended' in the glass, such as logos for example. When lit, light plays on the facets of the splinters, making them shine like crystals.

1

1 & 2 Guests are welcomed by an obelisk made of the Glaszone material at the May coffeehouse/cake shop in Neufeld, Austria.

© Rainer Zottele

2

Contact
Kolpa
Rosalnice 5
8330 Metlika
Slovenia
+386 (0)7 3692 100
kerrock@kolpa.si
www.kerrock.si

Objectile
www.objectilediffusion.com

Gouged Kerrock Lumino

Composition
Aluminium hydroxide (obtained from bauxite) and acrylic-based polymer binder

Properties
Decorative and continuously varying translucence

Applications
Light fittings, furnishings, wall coverings, etc.

Technical specifications
- Thicknesses: from 6 to 12 mm
- Properties similar to standard Kerrock (less hard and more easily thermoformable)
- Colours: white, pastel blue, yellow, pink and green
- Special adhesive for transparent colours

This material from the Solid Surfaces family has an acrylic and mineral base. There are numerous forms and colours and even a translucent version. These panels are the result of collaboration between Kolpa, the material manufacturer, and Objectile, the company making the products. Kerrock Lumino, once sculpted, reveals all its translucent qualities. The varying thickness of the material enhances the effect of light playing on it.

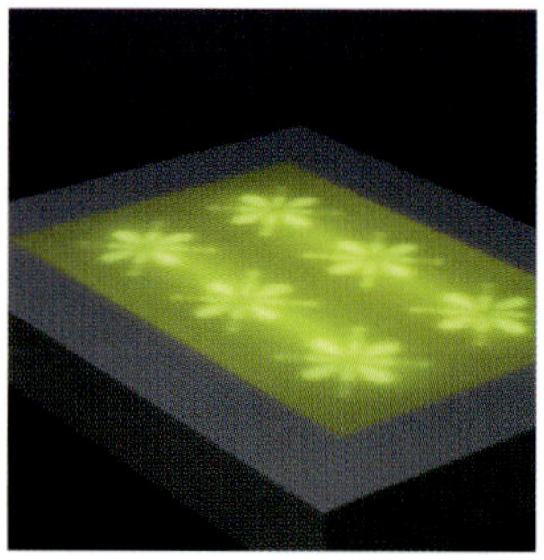

82

Contact
Graphic Concrete
Porkkalankatu 11 H 11
00180 Helsinki
Finland
+358 9 6842 0093
info@graphicconcrete.fi
www.graphicconcrete.fi

Graphic Concrete™

1

Graphic Concrete™ offers a new way to produce a patterned concrete surface. The patented Graphic Concrete™ technology involves the application of a surface retarder to the surface of a special membrane. This method allows precast concrete manufacturers to produce high-quality concrete elements and slabs. The end result is a patterned, smooth or completely exposed surface. The pattern on the surface results from the contrast between the fairface and the exposed fine aggregate finish. This is a cost-effective, tested and safe method because traditional exposed fine aggregate materials and processes are used in the manufacturing of concrete elements.

2

Composition
Concrete

Properties
Decorative, suitable for concrete elements and slabs of all sizes, stylish and reasonably priced, durable and easy to maintain, tested and safe, and custom designs possible

Applications
Façades, partitions, walls and slabs

Technical specifications
· Maximum width: 3200 mm
· Possibility for creation of customised motifs

1 For the 'Kampin Huoltotunneli' tunnel the city of Helsinki chose a tree pattern

2 For a residential building in Helsinki B&M Architects used 100 m² of façade cladding for an external wall panel

3 Frölunda Torg, Gothenburg, Sweden

3

84

Contact
Ecovative Design
1223 Peoples Avenue
Troy, NY, 12180
United States
+1 518 690 0399
info@ecovativedesign.com
www.ecovativedesign.com

Greensulate™

Composition
Natural byproducts (buckwheat hulls, cotton burrs, etc.)

Properties
Light, biodegradable, using renewable resources, insulating, firewall action, keeping homes efficient and safe, no allergy or health risks, and requires much less energy to produce than conventional rigid boards

Applications
Insulation

Technical specifications
- Thermal Conductivity: 0.06 W/mK
- Flammability: non-flammable
- Density: 96 kg/m^3
- Compressive strength: 1.193 MPa (ASTM C165-07)
- Flexural strength: 0.172 MPa (ASTM C203-05a)

Greensulate™ is an environmentally-friendly replacement for the expanded polystyrene structural insulating panels. Greensulate™ is produced from natural byproducts, such as rice hulls and cotton burrs (waste from cotton harvesting), without heat or light input. With high strength and excellent insulating properties, Greensulate™ insulation helps to reduce the amount of energy used to build houses and increases house 'green' efficiency with time. This product embodies cradle-to-cradle philosophy. Every step of Greensulate™ insulation's production, from the materials it's made from, to its manufacturing process and even to its end-of-use, is environmentally friendly. Greensulate™ is made from untapped byproducts local to the area where it is produced, conserving resources and minimising its environmental impact after use.

See notions: **Cradle to cradle™** p.218

86

Contact
Sander Irmãos
Rua Santo Agostinho 913
São Leopoldo 93025-700
Brazil
+51 (0)3589 6644
sander@sander.com.br
www.sander.com.br

Horn and Bone Craft

Composition
Horn and bone

Properties
Decorative and unique elements

Applications
Decoration, fashion, accessories (buttons, knife handles, etc.), jewellery, dog toys, etc.

Technical specifications
The techniques used in making products from these cattle-sourced materials are environmentally friendly. All production residues are recycled. Being non-polluting, they are generally used as fertiliser

This company specialises in processing horn and bone obtained from cattle. Each piece is unique and rich in detail. Horns are cut open and pressed flat to obtain elements which can be left in this raw state or polished as required. Bones are degreased and can be coloured before being machined. Among other applications, small horn and/or bone elements can be used in mosaic form to create decorative wall panels.

88

Contact

Marianne Kemp
Borneostraat 90huis
1094 CP Amsterdam
the Netherlands
+31 (0)6 1525 4840
info@horsehairweaving.com
www.horsehairweaving.com

Horsehair Weaving

Composition

Mongolian horsehair, linen, wool, silk and cotton

Properties

Decorative, light play, textures and colours, and handmade

Applications

Decorative wall panels, window panels, screens

Technical specifications

- Dimensions: 800 x 600 mm or 1500 x 1200 mm panels, other dimensions possible
- Creations are generally made-to-measure

Marianne Kemp weaves horsehair elegantly in an unconventional way, mixing in other fibres such as cotton or linen. She creates panels, unique handmade pieces, playing with the textures and colours of materials and raised patterns. The horsehair is woven, rolled, looped, etc. in order to play with its specific shiny properties and length. Some of her creations are now produced mechanically and are available through the John Boyd Textiles company in the United Kingdom.

1

1 Natural materials, wool and horsehair are combined into new textures with unexpected volume and dramatic curves. Here you see pink and green loops.

2 Knotted lines.

90

2

Contact
Hortus verde
55 quai de Valmy
75010 Paris
France
+33 (0)1 4201 2442
contact@hortusverde.com
www.hortusverde.com

Hortus Verde

1

Composition
Natural trees and plants

Properties
Decorative, no maintenance, no parasites, and no light necessary

Applications
Interior decoration

Technical specifications
Sensitive to humidity and excessive temperatures

The eternal garden. Hortus Verde has mastered the stabilisation of vegetation. Thanks to a secret 100% natural, magic formula, ageing of trees and plants is gently suppressed. This allows greenery to be introduced anywhere, in a restaurant, hospital, etc. without the need for light, with no maintenance and zero risk of parasite introduction. Vegetation is 'frozen' at its best.

2

92

3

1 A carpet of grass runs along the wall up to the ceiling.

2 A cascade of grass, a plant fresco to refresh and enliven a wall.

3 The art object 'levitating tree'.

4 A partition wall.

4

Contact

Lamera
Stena Center 1A
412 92 Gothenburg - Sweden
+46 (0)31 772 6095
info@lamera.se
www.lamera.se

94

Hybrix™

Composition

Stainless steel

Properties

Light, strong and recyclable

Applications

Aircraft and car interiors, light furniture, military use, absorbent anti-ballistic structures, etc.

Technical specifications

- Various dimensions available, common ones for example are:
Overall thickness 1.2 mm
Faceplates of 0.2 mm
Core 0.8 mm, consisting of stainless steel fibre 22 μm thick, distributed in density at 5%
- The thin faceplates on either side of the fibre can be textured, coloured, etc.

A stainless steel sheet, ultra light, consisting of millions of microscopic fibres, supported by two thin stainless steel faceplates. The product is of interest as, among other attributes, it offers a saving in material and weight. This sandwich construction is, in effect, 90-95% air. Furthermore, consisting of 100% stainless steel, it is recyclable, able to be melted down again much more easily than a composite sandwich material. The sheets can be cut, formed and machined in a similar manner to solid metal sheets and with the same tools.

1

1 Due to the high surface quality and low weight of Hybrix™, it's a very attractive material for portable consumer goods, like this embossed Hybrix™ stainless steel suitcase prototype.

Contact
Soejima Isao Shoten
23-1 Nakakimuro
831-0014 Ohkawa-shi Fukuoka
Japan
+81 (0)944 88 1141
info@soejima.com
www.soejima.com

Igusa

Composition
Igusa (Japanese rush, *Acorus Gramineus Ogon*)

Properties
Flexible, strong, light, air purifier, humidity regulator, insulating and soothing perfume

Applications
Japanese rush mats (tatamis), woven textiles, etc.

Technical specifications
Available in various colour ranges

The Igusa plant has been cultivated and recognised for centuries in Asia for its qualities of flexibility, strength and lightness. Once woven, the straw acquires a three-dimensional texture which has a massaging effect, stimulating blood circulation in the feet. Its fibres are aerated by means of countless natural perforations. The stalk thus has good insulating properties, retaining heat in winter and coolness in summer. The mats also provide some air purification and humidity regulation. The perfume of Igusa is very pleasant and has aromatherapeutic virtues. This long-stalked plant is however fragile and is not very resistant to wear or to sunlight.

1

2

1 Igusa floor mats are commonly used in Japanese interiors.

2 This lounge chair of the IGSA series is made of Igusa (Japanese rush). The designer dyed the material with natural products before mounting it on the metal frame.

Contact
Bader
Metzgerstrasse 32-34
73033 Göppingen
Germany
+49 (0)7 1616 7280
info@bader-leather.de
www.bader-leather.de

1 Injection moulding examples.

2 Sheets of Kollamat.

Kollamat

Composition
Leather fibre, biopolymers, collagen material and other additives

Properties
Great similarity to leather: warm to the touch, high aesthetic quality, characteristic smell of leather, sound and heat insulator, and variable density and strength

Applications
Cosmetic effects, use in automobiles, etc.

Technical specifications
· Used for high quality injection moulding with very good accuracy of shape, low energy use (max. 160ºC (320ºF)), biologically degradable, and ideal for components intended to feel and look like leather
· Based on polymers with low melting index: LDPE, PP, EVA, TPE, SEBS and also biopolymers
· Possible percentage of leather fibre: 5% to 70%
· Possible additives: mould release agents, lubricants, silicones, etc.

This compound is loaded with 100% natural leather fibre. The result, once transformed into a normal thermoplastic, offers all the aesthetic qualities of leather associated with the technical qualities of a polymer.
The typical characteristics of leather, such as being comfortable to the touch, are preserved. As for the product's environmental credentials, these are good as it makes use of scrap from the leather industry.

1

2

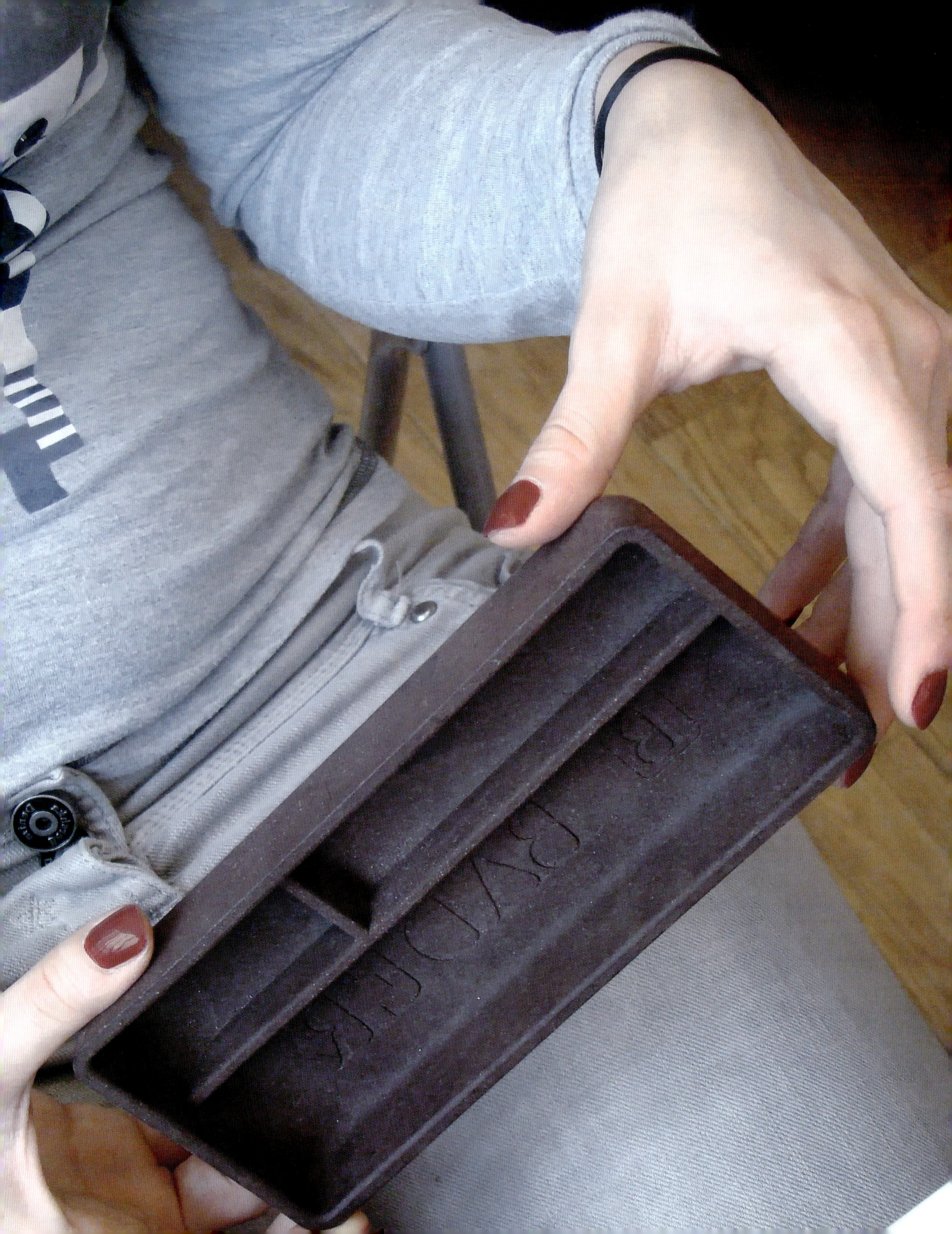

Contact
CWP - Coloured Wood Products
Jänhiäläntie 9
55300 Rauha
Finland
+358 (0)50 377 3766
info@cwp.fi
www.cwp.fi

Laminated Blanks

1

104

Composition
Birch wood

Properties
Decorative, excellent alternative to paints and lacquers for products subject to high wear and tear or outdoor use, custom colouring and effects

Applications
Hunting and sports-related equipment, design products, interior design and furnishing solutions, as well as applications in the building industry

Technical specifications
The standard size of a blank is 820 x 280 mm and they are available ex-stock in six different standard colours

The speciality of this Finnish company is self-coloured birch, produced by a method which allows pieces of wood up to 45 mm thick to be coloured throughout to an exact desired colour. The operation can be repeated to give identical results (for a particular consignment of wood). Once coloured, veneers can be combined and assembled to reconstitute solid, easily-machined panels, whose colour shades are fully revealed in 3D shapes.

3

1 & 2 Birch veneer coloured throughout, is ideally suited for various kinds of gifts, like these plates and balls.

3 The Kurvinen table has a plain multi-laminar top and curved legs. The uniqueness of the material is emphasised in the three-dimensional shape of the legs.

4 Scandinavian birch is a tough and strong hardwood, the colours and glues are suitable for outdoor use and therefore the perfect material for hunting gear.

2

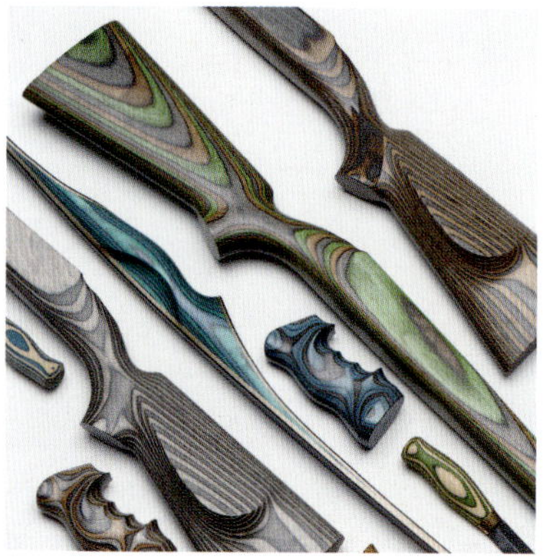

4

Contact
BBS Consult
3 allée de la grande Egalonne
35740 Pacé
France
www.bbsconsult.fr

Laser Engraving

Composition
A laser can be used to etch various substrates such as metal, stone, plastic, glass, etc.

Properties
Decorative, delicacy of patterns and high precision

Applications
Traceability, signage and decoration

106

Laser technology allows high precision engraving on a wide variety of different substrates. Laser action ranges from ablation of thin layers to colouring or full hollowing out of material. A technique which allows progressive 'silvering', very fine patterns, precise cutting to be obtained.

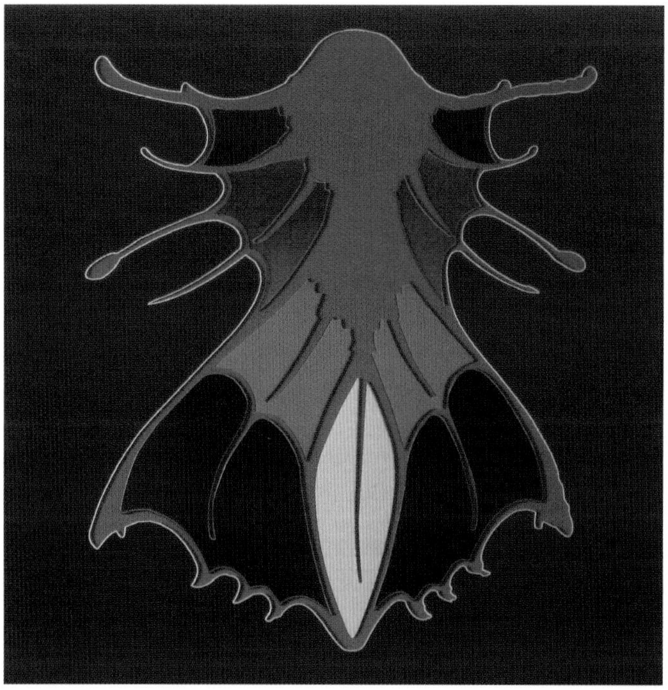

1

1 Stainless steel. The Physical Vapor Deposition (PVD) layer is engraved and the colour of some areas are modified.

Contact
Damglass Innova
Benavent 10
08028 Barcelona
Spain
+34 (0)93 3399 683
innova@damglass.com
www.damglass.com

Lemonpack
www.lemonpack.com

Composition
Glass brick

Properties
Decorative and sound insulating

Applications
Partitions

Technical specifications
These bricks are laid in the same way as normal glass bricks

Lemonpack

Just imagine a brick made of glass in which you could include all sorts of materials, any such content being bathed in a liquid which, on the one hand, improves transparency and on the other protects against bacterial problems and prevents condensation. An original concept to provide your environment with visual effects limited solely by your imagination. Lemonpack is handmade and can be customised in any way the customer desires. An additional benefit: any liquids and objects included give the brick more weight and provide better sound insulation for between partitioned areas.

108

Contact
Bencore
Via S. Colombano 9
54100 Massa
Italy
+39 05 8583 0129
info@bencore.it
www.bencore.it

Madreperla
www.madreperlaspa.com

Lightben Kaos

The sandwich panel, commonly known as 'honeycomb panel', breaks with its predecessors by offering randomly irregular cavities. In effect, Lightben Kaos is composed of a polycarbonate core with variable-diameter cylindrical cells, assembled with external acrylic walls. The resulting effect gives a pleasing, chaotic effect, moving away from the industrially-produced, uniform cell structure towards a more natural, random effect.

Composition
Polycarbonate honeycomb core, and acrylic skins

Properties
Translucent, decorative, resistant to compression, lightweight, stiff and suitable for back lighting

Applications
Indoor furniture, display counters, partitions walls, doors, etc.

Technical specifications
- Thickness: 19 - 21 mm
- Width: 1000 mm
- Length: 3015 mm
- Solely for indoor use
- Easy to cut
- Partitions available in various finishes and colours
- Can be finished with standard accessories profiles and systems such as frames, handles and hinges
- Can also be edged with the same material used for the external layers

110

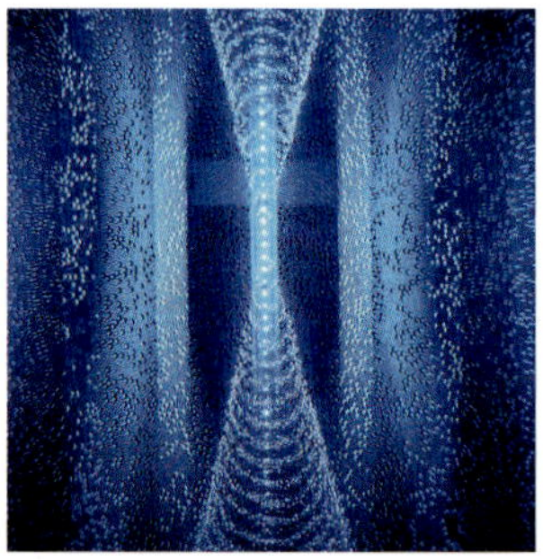

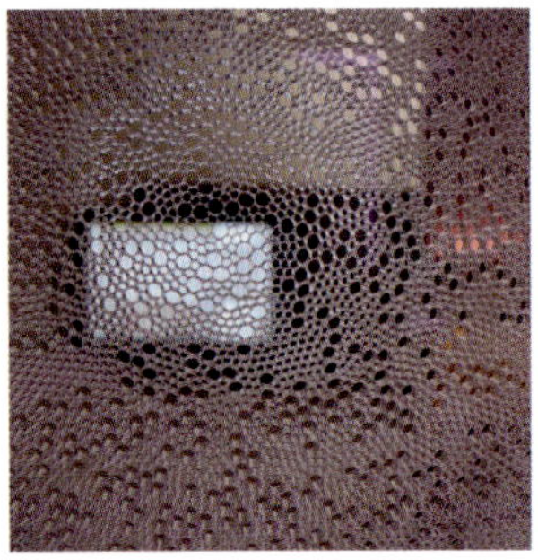

Contact
Brochier Technologies
90 rue Frédéric Faÿs
69100 Villeurbanne
France
www.brochiertechnologies.com

Lightex®

Composition
Optical fibre, LEDs and textile

Properties
Flexibility, optical effects and decorative

Applications
Lighting, signage (text, pictograms and other communication functions), transport, architecture, light therapy, photocatalysis, etc.

Technical specifications
· Monochrome or RGB light
· Low voltage
· Low power consumption
· Lifespan: 50 k to 100 k hours
· Textile areas up to several tens of square metres
· Bright surface is ultra-thin (<1 mm) and ultra-lightweight
· Surfaces can be flexible, rigid, flat, curved, 3D
· Patented process

Principle: End-lit, woven optical fibre. LEDs connected to the fibres reveal patterns or illuminated surfaces. These luminous, flexible textiles (which can possibly be made rigid) take up little space, have low electricity consumption and are long-life. Of interest in various fields: new light sources for transport, luminous glass partitions in architecture, diffused light for phototherapy used in the treatment of sleep problems, pollution abatement by photocatalysis.

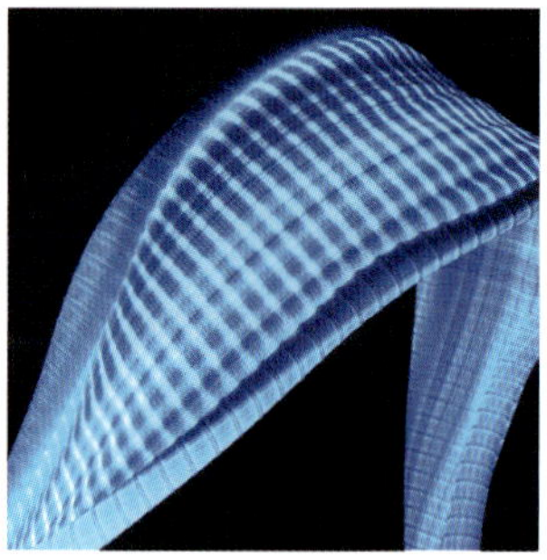
© Aurélie Foussart

112

Contact
Luccon Lichtbeton (HQ)
Bundesstrasse 1
6833 Klaus/Vorarlberg
Austria
office@luccon.com
www.luccon.de

Luccon Veneer

Composition
Concrete and quartz optical fibre

Properties
Translucent, fine thickness, decorative and flexible

Applications
Automotive and yacht engineering, room furnishings and décor, etc.

Technical specifications
- Maximum dimensions: 100x50cm
- Thickness: 0.8mm
- A film sub-layer allows 3D shape creation
- High UV resistance
- Veneers can be cut, glued and perforated

Luccon translucent concrete, consisting of layers of concrete and optical fibre, is now available in thin flexible sheets. For such a thin layer of material, these novel veneers offer a concrete effect, translucence and an astonishing 'mineral' appearance.

1

1 & 2 A study for internal car parts.

118

2

Contact
Litwork (HQ)
Schweizerstrasse 35
6844 Altach
Austria
+43 (0)5 5762 0852
office@luminoso.at
www.luminoso.at

Luminoso

Luminoso is a composite wood panel, comprising alternating thin layers of solid wood and optical fibre, the whole assembly securely bonded together. The optical fibres allow light conduction, thus producing a translucent wooden panel. The strength of Luminoso is equal to the strength of the chosen wood species.

Composition
Wood and optical fibre

Properties
Decorative, translucent and strong

Applications
Interior architecture (floors, walls and ceilings), discreet signage, nautical fittings, cars, aircraft, etc.

Technical specifications
· Dimensions: 3000x500mm and 3000x1000mm
· Thickness: 10, 12, 14, 16, 18, 20, 25, 30, 35, 40, 45, 50mm and other thicknesses on request
· Thickness of light conductors: standard 0.25mm
· Spacing between light conductors: standard 5mm and other thicknesses and spacing on request
· Wood species: Oak (*Quercus* - many varieties), Cherry (*Prunus* - more than one variety), Walnut (*Juglans* - more than one variety), Maple (*Acer* - more than one variety), Sipo (*Entandrophragma utile*), Teak (*Tectona grandis*) and Wengé (*Millettia laurentii*)

1

2

3

1 The new Luminoso film technique makes it possible to create any image on your Luminoso wall.

2 A bar counter with different motifs on it.

3 Depending on the available fibre optical waveguide, the wood material enables the ingression of daylight and direct sunlight. Objects, images and shapes being placed between the light irradiation and Luminoso, block the light and turn into creative shadow plays.

Contact
Absel
1 square La Bruyère
75009 Paris
France
+33 (0)1 4526 0339
www.absel.fr

Lumisheet

Composition
LEDs and acrylic diffusing plate

Properties
Very thin, light, easy to install, waterproof, long life, zero maintenance, low power consumption, flexible in shapes, economical to integrate and high quality

Applications
Suitable for interior and exterior use. Furniture, interior architecture, etc.

Technical specifications
· Dimensions: from 60 x 152 mm to 600 x 1800 mm, other sizes on request
· Thickness: 4 - 8 mm

Lumisheet is a lighting system which integrates LEDs directly in a 3D diffuser plate. The light produced is uniform and the slenderness of the device opens up numerous applications.

122

Contact
Inn'deko
12 rue Raymond Barbet
92000 Nanterre
France
+33 (0)6 1089 9177
cb@inndeko.com
www.inndeko.com

Metalis

Composition
Metal

Properties
Decorative, excellent adhesion and high resistance to bending

Applications
Wall coverings and furniture

Technical specifications
Metalis powder can be applied with trowel, float, spreading spatula, paint brush or by spraying

124

Based on an aqueous solution, this technique allows cold metallisation on any substrate, flexible or rigid. Being very easy to implement, the effect can be applied to many different surfaces and allows multiple solutions for wall finishes.

1

2

3

1 A sample of different available colours.

2 Detail of a windowsill.

3 A metal zinc-look is applied to this concrete stairs.

4 A door treated with the Metalis material.

4

Contact
BASF
www.basf.de

Mincor® PES

Composition
A compound containing nanoparticles and a water- and oil-repellent resin

Properties
Self-cleaning

Applications
Awnings, sunshades, sails, tents and fashion industry

Technical specifications
The feel of a textile remains unchanged, different types of textiles can be treated, and no dispersion of nanoparticles

It is possible to reproduce the self-cleaning ability of the lotus plant by treating a textile composed of synthetic fibres with Mincor® PES (covers textile with the a nano porous layer). It so happens that textured surfaces rather than the smoothest surfaces are the most effective at repelling water and dirt. Dirt is not able to stick and is carried away by water droplets which form on the treated fabric. This self-cleaning property is invaluable in the case of textiles used outdoors, as rain has a self-cleaning effect. Awnings for example, for which machine washing is impossible, can stay 'new-looking' for longer.
According to current knowledge there is no exposure to these nanoparticles during the production and use of Mincor® PES. Mincor® PES complies with the strict rules of BASF code for nanotechnology.

See notions: **Biomimetics** p.217, **Lotus effect** p.225 and **Nanoparticles** p.226

Contact
Salmo Leather
Birkenthal 8
94253 Bischofsmais
Germany
+49 (0)9920 9400 55
info@nanai.com
www.nanai.eu

Nanai

1

This product benefits from a chrome-free vegetable tanning technique, which has the advantage of maximum preservation of the structure and pigments of the salmon skin. Based on a Siberian tanning tradition and combined with new technologies, this tanning method maximises the strength and durability of the skin. The salmons come from certificated organic farms in Irish aquaculture.

2

Composition
Salmon-skin leather

Properties
Chrome-free vegetable tanning, strong, durable, light, thin and tear-resistant

Applications
Textile, footwear, jewellery, decoration, furnishing, etc.

Technical specifications
· Skin length: 550 to 650 mm
· Maximum skin width: 170 mm
· Yard goods: rectangles (8 cm x 40 cm) are cut out of the skins and sewed
· Large choice of possible colours
· Rain-resistant
· Very hypoallergenic, non-irritating to the skin

3

1 During the Berlin Fashion Week a model showed a skirt made of Nanai Salmon skin designed by Michael Michalsky.

2 A handbag by Michael Michalsky.

3 Nanai is not only being used as a decorative wall cladding in residential interiors as shown here, but also in commercial interiors.

4 Nanai is being used extensively for upholstery.

4

Contact
GKD (HQ)
Metallweberstrasse 46
52353 Düren
Germany
+49 (0)24 218 030
info@gkd.de
www.gkd.eu

Native Bamboo

Composition
Bamboo and stainless steel

Properties
Decorative, weather- and corrosion-resistant and sturdy

Applications
Suitable for interior and exterior use: partitions, façades, handrails, etc.

Technical specifications
· Bamboo diameter: 26 - 28 mm
· Thickness: 30 mm
· Open surface between 25% and 39% according to references
· Weight: between 7 and 10 kg/m^2 according to references
· Maximum length: 15 m
· Maximum width: 3 m

These materials quite literally interweave the natural beauty of bamboo with the strength of stainless steel. Bamboo rods are integrated with metal materials whose distribution of rods and cables are variable, thus creating multiple possibilities for variation.

Contact
Michel Graveur
Zone d'activité
69170 Valsonne
France
+33 (0)4 7405 1680
michel@michelgraveur.fr
www.michelgraveur.fr

Nickel Plating

Composition
Nickel

Properties
Decorative, high cutting precision, and therefore reproduction of images, and transparency

Applications
Interior architecture, furniture, light fittings, etc.

Technical specifications
· Maximum dimensions: 3000 x 800 mm
· Thickness: from 70 to 190 μm as a function of use

This technology was initially used for printing motifs on textiles via a cylinder. Today, it is finding new applications in architecture, design and decoration. Thousands of micro-perforations are made on nickel plates or cylinders in order to reproduce a design or an image. The sparkle of the metal foil and the transparency added by the perforations allow light to play elegantly over surfaces.

Pearlnera®

This company specialises in worktops, sinks, counters, furniture or floor slabs in resin-mineral materials. The Pearlnera® product is a mixture which includes glass balls of different sizes. The inclusion of glass balls gives exceptional 'depth' to the material, an intriguing 3D effect. The appearance varies according to the angle of light incidence. The glass balls can be replaced by fragments of fossil, shell, etc. This gives an opportunity to personalise furnishing elements in colour and texture.

Contact
material raum form
Wielandstrasse 3
22089 Hamburg
Germany
+49 (0)40 4857 02
www.material-raum-form.com

Composition
Glass balls and resin loaded with mineral

Properties
Decorative, solid and optical 'depth' effect

Applications
Interior floors and walls, as well as worktops, counters, basins, etc.

Technical specifications
- Maximum length: 4100 mm
- Emission-free
- The surface is generally polished and therefore smooth, revealing the glass balls. A rougher finish is possible
- Several colours are available and any colour is available on request

1

2

1 The glass spherules contained in the material of this wash-basin continually create new effects, depending on the incidence of light at different times.

2 A finely polished tabletop with a seamless surface. There is a choice between standard colours and dyes blended especially for the customer.

146

Contact
Dickson Constant
10 rue des Châteaux
59443 Wasquehal Cedex
France
+33 (0)3 2045 5959
contact@dickson-constant.com
www.dickson-constant.com

Photovoltaic Textile

Composition
Textile

Properties
Light, flexible and strong

Applications
Awnings, fashion industry, equipment for sports, military use, camping, etc.

Technical specifications
· Life of PV cells: approximately 20 years
· Efficiency 5%

This fabric gives personal protection against the sun, rolls up and produces electricity. The principle? Lightweight (900 g/m^2) and ultra-thin (<1 mm) photovoltaic cells are integrated directly into awning fabric. The cells, of the amorphous silicon type (a-Si), are also flexible, allowing an awning to be rolled up. When photons hit the surface of these cells, their energy is transferred to the electrons of the silicon. The energised electrons create an electric current which is fed into the local power network via an inverter or converter. The electricity generated can then be used to power equipment.

148

Contact
FRED&FRED
info@fred-fred.com
www.fred-fred.com

PICT®

1

Composition
Glass brick

Properties
Decorative, optical effect,
easy to lay and maintain and fully recyclable

Applications
Interior partitions

Technical specifications
· Dimensions: 225 x 225
· Thickness: 90 mm

This patented glass-brick system plays with optical effects, breaking an image down into several small colour transparencies. These modules and their 16 lenses transmit light and images, while preserving the intimacy of the spaces enclosed. In addition, this optical-glass system has an integral system to eliminate any thermal bridging.
There are four variants: Original PICT®, Frosted PICT®, Transparency PICT® and Black PICT®. PICT® is a brand deposed by the company FRED&FRED.

2

1 The manufacturing process.

2 Black PICT®.

3 A red version of PICT® developed for Jean Nouvel's Red Sun Pavilion, Serpentine Gallery, London.

4 The optical glass wall system transmits both light and images while preserving the privacy of the walled spaces.

3

150

4

Contact
Bleu Nature (HQ)
26 rue Favreuil
59100 Roubaix
France
+33 (0)3 2011 2528
contact@bleunature.com
www.bleunature.com

'Pixillated' Wood

1

Composition
Wood from certified, sustainably-managed European forests

Properties
Decorative, easy to cut, sound-insulating and water-repellent (on request)

Applications
Wall coverings

Technical specifications
· Dimensions: 750 x 1150 mm and other dimensions on request
· Thickness: 68 mm
· Weight: 26.5 kg/m²
· 3 variations available: Pixels of driftwood in relief, Pixels of cut driftwood in relief and Pixels of pierced driftwood in relief

This rough wall covering, with the air of a mountain chalet, consists of dozens of slices of small wood logs, randomly juxtaposed. These grouped pieces of wood are glued onto a sheet of oriented-strand board (OSB) with a rough-cut finish (could also be polished or pierced). Tree species are mixed and untreated, showing all the fascinating variations of colour and texture in the saw-cut, non-sanded wood. Each panel is made in the North of the France (Roubaix) and each piece of wood is positioned by hand during manufacture.
Another material available from this company consists of mounted sections of petrified wood. Petrified wood requires no design, variation or arrangement. Its simple presence is enough.

2

1 & 2 For wall and floor coverings a collection of wood pieces are glued vertically to particle board tiles.

PURE

Contact
Columbia Forest Products
7900 Triad Center Drive
Suite 200
Greensboro, NC, 27409
United States
+1 (0)800 637 1609
www.columbiaforestproducts.com

Pure Bond® Hardwood Plywood

Composition
Soy-based formaldehyde-free adhesive, hardwood plywood

Properties
Non-toxic, competitive price and water resistant (more than traditional formaldehyde adhesives)

Applications
Building, furniture, etc.

Technical specifications
· Panels can be specified as FSC-certified upon request
· Panels are LEED® compliant and may contribute points for EQ Credit 4.4 for low-emitting materials: Composite Wood.
· Manufactured exclusively in North America

By replacing the traditional adhesive used in the manufacture of plywood panels by a non-toxic adhesive, this company has eliminated any added formaldehyde (urea formaldehyde, UF) from its wood-derived products. Inspired by the adhesion abilities of the mussel and the protein which assures this function in that animal, this adhesive is synthesised from soya. A good example of biomimetics in action.

See notions: **Biomimetics** p.217 and **FSC** p.223

162

Contact
Canjaere
Route de Ham-les-Moines
08090 Tournes
France
+33 (0)3 2452 9320
infos@canjaere.fr
www.canjaere.fr

Querex

164

Composition
Oak from French forests

Properties
Decorative, strong and low-maintenance

Applications
Outdoor terraces, paths, balconies, etc.

Technical specifications
- Dimensions: 120 x 120 mm
- Thickness: 60, 80 or 100 mm
- Products are stabilised with linseed oil

In the past, before the use of bitumen as a top coating for roads and pavements, traffic areas of Paris were covered with wooden setts. These slabs have not completely disappeared. Some traces remain in certain passageways and this company continues to make the material. The oak surface is now used for external terraces, for example. The setts are bedded on sand, after ground stabilisation according to the anticipated loading. This material offers a wide variety of possibilities for motifs.

Contact

Tamiya raden textile
312, Miyake Tango-cho
Kyotango-city
627-0121Kyoto
Japan
tamiya@gold.email.ne.jp

Raden Textile

Composition
Nacre (mother-of-pearl) and paper

Properties
Decorative, flexible, iridescent and expensive

Applications
Clothing, furnishings, etc.

Technical specifications
· Complete or partial mother-of-pearl weaving (motifs, etc.)
· A very expensive product

This ancient technique involves the extraction of fine layers of nacre from certain shellfish. Thin pieces are then glued to paper, this also being cut into thin pieces which are finally woven together. The result is astonishing: a flexible material, of real nacre, on which the light plays quite beautifully.

166

Contact

Raincampagn
World Fashion Centre
Koningin Wilhelminaplein 13
Tower 1, floor 13
1062 HH Amsterdam
the Netherlands
+31 (0)20 6152 429
info@raincampaign.com
www.raincampaign.com

Raincampaign

Composition
Concrete

Properties
Magical and hard wearing

Applications
Signage, publicity and architecture

Technical specifications
- Non-noxious
- Effect lasts for at least 8 weeks

168

Raincampaign is a hydrophobic surface treatment applied to concrete which, once wet, reveals a pattern, message, etc. A good way to display publicity messages in an environmentally friendly way.

1

2

3

1 Raincampaign developed the holiday advertising campaign for the Dutch travel agency D-Reizen. Every time it rains the message: 'Are you also yearning for the sun?' appears on the pavement. The locations are carefully positioned in close proximity to the D-Reizen stores.

2 Every time it rains the campaign advertisement reappears on the pavement. When the tiles dry up the Raincampaign disappears again. The message will be visible on the pavement for at least 4-6 weeks.

3 The winter tyres from the Continental tyre company work better in rainy conditions. At 500 locations in the Netherlands the message: 'Stop faster with Continental winter tyres' appears when it rains. The locations are carefully positioned in the areas of car parks, car park ticket machines and crowded places.

© Courtesy of Raincampaign

Contact

Grado Zero Espace
Via nove 2/A
50056 Montelupo Fiorentino
Italy
+39 05 7180 368
contact@gzespace.com
www.gzespace.com
www.fabriclightbulb.com
www.tessutolampadina.it

ReLight

1

Composition
Fabric made of glass fibres, electro-active nanopolymers and phosphor

Properties
Light, flexible, durable, resistant to extreme conditions, perforation, abrasion, torsion and compression, flame resistant, and doesn't contain mercury or other toxic substances

Applications
Fashion industry, industrial design, military, sportswear, security/safety, architecture, etc.

Technical specifications
· Any type of textile can be used as a substrate, as well as rigid materials such as glass or metals
· Colours: white, yellow, blue, red and green
· Power requirement: 25-50 W
· Light efficiency: between 1.5 and 4 lumens/W
· Internally reflected component (IRC): between 73 and 84
· Panel weight (800 x 1000 mm): 300 g
· Gelbo Flex Test (no. of cycles): passed (400 cycles)
· Flame Test, ASTM D6413-99 (Standard Test Method for Flame Resistance of Textiles - Vertical Test): passed
· EMI Interference, MIL-STD-461E (Requirements for the Control of Electromagnetic Interference Characteristics of Subsystems and Equipment): passed

ReLight is a new-generation light, based on electroluminescence technology. First developed exclusively for the military sector by a company at the forefront of research on electro-active polymers, this material is composed of glass fibres with micro-coatings of conductive polymers. This is therefore a cold-light source which is flexible and presents no danger for a person handling it. The light emission is diffuse and uniform and opens the way to various applications where the ability to 'fold' the light, for example for transport or packaging, may be very valuable.

See notions: **Electroluminescence** p.220

1 In the dark the seats in this sports car come alive thanks to the azure blue ReLight fabric between the carbon fibre frame and the Techno-gel cushion which gives the driver and passengers the spectacular impression of an endless space.
© Courtesy of Bertone

2 ReLight panels are pre-installed onto the US Army tent liner to enable fast deployment of the air beam tent when needed.
© Courtesy of US Army

3 A 170 cm² ReLight panel undergoing the Gelbo flexibility test. ReLight panels can withstand up to 100 twist cycles with minimal degradation.

2

170

3

DURACELL

Contact
Arkema
420 rue d'Estienne d'Orves
92705 Colombes Cedex
France
+33 (0)1 4900 8080
www.arkema.com

Reverlink™

Composition
Semi-crystalline resins, self-healing elastomers, supramolecular modifiers ('stickers') and adhesion promotors

Properties
Self-healing, bio-sourced products, adhesion on plastics, metals (anti-corrosion), glass, outstanding energy dampening ability in a wide temperature range (amorphous grades), UV resistance and electric insulation, material recycling enabling

Applications
Conveyor belts, gaskets, anti-impact protection, shock-absorbing layers, professional gloves, fast closing septa, anti-corrosion coatings for metals, additives for various formulations (adhesives, bitumens, organic binders, paints, varnishes, pastes and mastics), etc.

Technical specifications
The Reverlink™ range currently comprises about ten grades including supra-molecular elastomer materials with optimum self-healing quality, products for modification of traditional polymers and different specialised additives. Can be formulated and blended with usual rubbers or thermoplastics, for new self-repairing and adhesion properties

These resins, elastomers, modifiers and promotors, with low temperature processability (semi-crystalline grades) and remarkable self-healing properties (rubber grades), involve supra-molecular chemistry and are derived from renewable raw materials. As a matter of fact they are based on fatty-acid oligomers issued from vegetable oils, representing 60% at least of their composition.
A feature of supra-molecular materials is that they have so-called 'reversible' (non-permanent) inter-molecular links, unlike the polymers arising from traditional chemistry, which rely on 'irreversible' (permanent) links. This reversibility characteristic gives rise to a whole family of high-performance materials which are distinguished by their complete-break properties in comparison with traditional materials, among others this amazing self-healing ability: a crack or break occurring in these materials can be repaired by simple contact between the opposing surfaces of the break under light pressure, recovering most of the initial strength. The alternative would otherwise be adhesive or heat. This property is thought to be of particular interest in composite materials which could last longer thanks to micro-cracks repairs or to a better cohesion between fibers or fillers and organic binders.

172

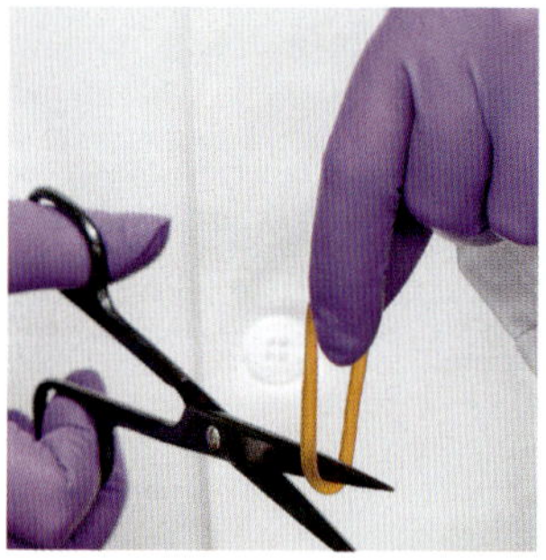

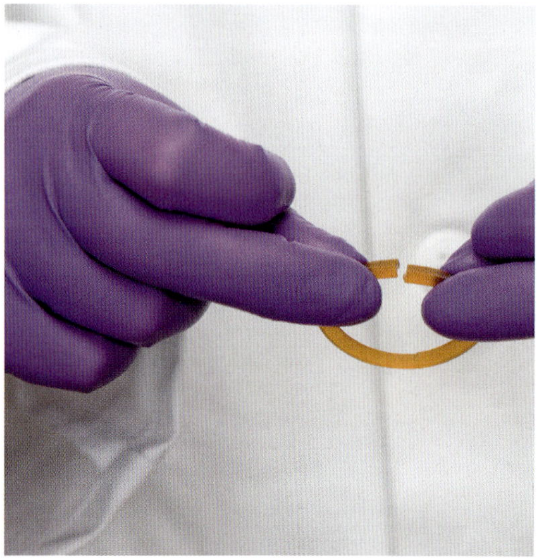

Contact
Atelier Jean Daniel Gary
100 rue Raspail
94700 Maison-Alfort
France
+33 (0)1 4378 4013
jdgary@club-internet.fr
www.jdgary.fr

Semi-mirror Glass

Composition
Glass and silvering

Properties
Decorative and interplay of translucency and reflection

Applications
Interior architecture

Technical specifications
· Handmade
· Any possible shape

Jean-Daniel Gary works his glass with great skill. Among other things, he sands, sculpts and colours it to create pieces with truly astonishing relief effects. He is currently mastering a new technique, a veritable tour de force: progressive silvering on float glass. A great way of toying with different magical facets of glass: its perfect transparency and then its totally reflective properties when it acts as a mirror.

See notions: **Float glass** p.222

174

Contact
SmartGlass International
21 Cookstown Industrial Estate
Tallaght, Dublin 24
Ireland
+353 (0)1 4629 945
info@smartglassinternational.com
www.smartglassinternational.com

SmartGlass™

Composition
Glass and liquid crystals

Properties
Adjustment of the transparency of laminated glass

Applications
Architecture and interior architecture (partitions)

Technical specifications
· Power consumption: 5 W/m²/hour and less than 200 mA/m²
· Power supply: 110 V AC

LC SmartGlass™ and SPD SmartGlass™ are two advanced laminated products which allow control of glass transparency, light, heat and glare.
Both types are based on polymer technologies, with liquid-crystal molecules or suspended particle devices in suspension. The LC type contains spherical particles which are aligned (transparent state) when the power is on and when the power is switched off there is an instant change to a state of random orientation (opaque state). This type is mainly used in hotels, office buildings and hospitals to provide privacy, where traditionally blinds would have been used.
The SPD type contains rod shaped particles which can be oriented by the user through applying varied voltage. This type provides control of glare, light and heat caused by sunlight, thus allowing users to control their working or living enviroment. Advantages include reduced lighting and air conditioning costs through day light harvesting. Both SmartGlass™ types block almost 100% of damaging UV.

See notions: **Liquid crystals** p.224

1

2

3

4

1 & 2 SPD SmartGlass™, electronically switchable solar control glazing, is used for a roof window at the Deparma showroom in London.

3 & 4 For the boardroom at Damico Tankers, the doors with LC SmartGlass™ make it possible to make the doors look see-through or not.

Contact

Nathan Allan Glass Studios
110 – 12011 Riverside Way
Richmond, BC, V6W 1K6
Canada
+1 604 277 8533
www.nathanallan.com

Stax

1

Composition

Glass and mirror

Properties

Decorative effect for large surfaces, stable and strong

Applications

Interior and exterior architecture, and façades

Technical specifications

- 3 models available: Beamz (straight layers), Breez (wavy layers) and Flex (twisted layers)
- Maximum dimensions of sheets: 1220 x 2750 mm
- Thickness of glass generally 32 mm but can vary from 25 to 38 mm
- Currently available as clear glass. Colour finishes applied to the rear face, with or without mirror treatment (which accentuates the strata)
- Weight: about 7 g/cm^2. A 1524 x 2743 mm panel weighs about 296 kg, and is therefore very difficult to manipulate

The Stax range offers an ingenious and functional way of creating external walls on very large areas. Several layers of glass are assembled and merged in layers to create solid, strong panels. The strata effects can be used vertically or horizontally.

2

3

1 & 2 Colour samples of the Stax glass.

3 Panels of Stax glass are seamed together to form a continuous façade along the front of this bar at Simon Restaurant & Lounge in New York. The distinctive ambience of Stax, Beamz pattern is enhanced with an amber colour finish and mirrored backing.

© Jeffrey Green Photography

Contact
Sto
Ehrenbachstrasse 1
79780 Stühlingen
Germany
+49 (0)77 4457 1010
infoservice@stoeu.com
www.sto.de

Stolotusan

Composition
Acrylic-base paint

Properties
Very hydrophobic, repels dirt, excellent resistance to weather and UV, and highly permeable to water vapour

Applications
External façades

Technical specifications
· Available in almost any colour, including customised colours
· Resistant to mould, mildew and algal growth

As time goes by, external façades tend to become dirty. Inspired by the lotus effect, where the flow of falling rain droplets makes it simply impossible for dirt particles to obtain a hold on the leaves of the plant, this external coating is extremely water-repellent. Its micro-structure has been based on the lotus in order to reduce the surface area in contact with water and dirt to a minimum. A façade therefore remains dry and clean.

See notions : **Biomimetics** p.217 and **Lotus effect** p.225

1

3

2

4

1 Stolotusan was used for the external façade at the Practicon school in Doetinchem, the Netherlands.
© Norbert Duijvelshoff

2 House Allegria in Tunbridge Wells, United Kingdom.
© Paul Freeman

3 The Lumenart office building is situated in Pula, Croatia. Intense whiteness dominates the interior and exterior as the background for the light design.
© A. Kaunat

4 Villa Tia by architect Andrea Mannocci.

184

Contact
Tellus-Ceram
Rue Beausoleil
47500 Monsempron-Libos
France
+33 (0)5 5371 9168
contact@terracruda.fr
www.terracruda.fr

Terracruda

Composition
Unfired clay

Properties
Solid, unbaked, zero maintenance, thermal and hygrometric regulation

Applications
Construction, decoration, etc.

Technical specifications
· Different formats, colours and motifs are available
· Dimensions: 230 x 76 x 114 mm (small bricks), 230 x 90 x 114 mm (large bricks), 230 x 172 x 114 mm (building blocks) and 305 x 230 x 114 mm (large blocks)
· No finishing operations necessary
· No maintenance
· Can be plastered
· Density: from 2 to 2.3 g/cm^3
· Mechanical strength equivalent to fired clay brick
· No shrinkage or cracking in a dry atmosphere, or swelling in a saturated atmosphere
· Recyclable material

These clay bricks or blocks are obtained by compacting the material under high pressure and without baking. Production here is semi-industrial and brings an ancestral material back into fashion. The compressed elements are solid and easily laid in the traditional manner. In addition, raw clay offers thermal comfort qualities (natural air conditioning by storing heat and releasing it later), hygrometric comfort (a breathing wall: high permeability to raised atmospheric water level and high capacity for absorption of damp) and acoustic comfort (constituting a sound screen).

186

Contact
Woodloop
Yliopistonkatu 2
20100 Turku
Finland
+358 (0)2 432 1270
woodloop@woodloop.fi
www.woodloop.fi

Textile Wood

Composition
Solid Finnish common aspen

Properties
Decorative, acoustic insulation and 100% wood

Applications
Interior wall covering and partitions, residential or public spaces

Technical specifications
· Dimensions: 14 x 50 x 1200 mm, 14 x 100 x 1200 mm, other sizes on request
· Thickness: approx. 5 to 9 mm
· Surface treatment: wax (carnauba / beeswax with sunflower / linseed oil)
· Easy maintenance
· Several colours available
· Fireproofing on request

Textile Wood is made from solid panels of common aspen (*Populus tremula*). High-pressure water is sprayed onto the panel surface, partially-detached fibres then form a 'moquette' of wood fibre, soft to the touch and with good sound absorbing qualities. The collection includes several different types: väre, aarni, aava, frans, viivo or tilkku, which exploit the textile effect in various forms: as different coloured bands, as a homogenous panel, as alternating bands of wood and defibrated wood, single-sided or double-sided, etc.

1

2

1 At the Information Media Technology Centre in Mikkeli, Finland, a textile wood with vegetable-based oil colours was used for a foyer installation.

2 A textile wood panel in a meeting room of Kinnarps Ltd., Helsinki, Finland.

Contact
Taporo
260, Boulevard Saint Germain
75007 Paris
France
+33 (0)1 5359 9489
contact@taporo.com
www.taporo.com

Textured Concrete

1

Composition
Ductal® concrete

Properties
Decorative and many possible surface effects

Applications
Interior architecture, and interior and exterior fittings

Technical specifications
Standard choice of textures but the company would rather work to order, large dimensions possible, and small production runs possible

Items are made using Ductal® concrete, which is loaded with very fine-textured glass fibre. This material allows great detail in the surface finishes: Arabian designs, crocodile 'leather', bubbles and even gold-leaf inlay, etc. All of these in muted colours of black, grey, dark green or chocolate, etc. The result is especially stylish. This company can also produce complete sets of furniture.

See notions: **Fibre-reinforced concretes** p.221

1 Sample of concrete texture.

2 Kitchen countertop, Ductal® concrete.

3 Wash-basin Earthshake.

4 Ecclesiastical furniture, Ductal® concrete.

2

3

4

Contact
Création Baumann
Bern-Zürich-Strasse 23
4900 Langenthal
Switzerland
+41 (0)62 919 6262
mail@creationbaumann.com
www.creationbaumann.com

Textile Gecko

This textile has one silicone-coated face which allows it to adhere easily to smooth surfaces such as glass. Rather like a 'post-it' textile, it comes off just as easily without leaving any trace. It will simultaneously provide protection against light and decorate windows, glazed partitions or doors.

Composition
Polyester PL or PLF CS and silicone

Properties
Decorative, provides protection from light, and easy to apply and remove

Applications
Shop windows, point-of-sale advertising, events, publicity, interior architecture, etc.

Technical specifications
· Large choice of colours, motifs and perforation patterns
· No fixing device or other arrangements necessary
· Machine washable
· Flame retardant

1

2

3

1 Gecko Crypta has a playful, graphic print design with hand-drawn, irregular loops and curls.

2 This bright orange colour is from the Gecko Crena collection.

3 Every form can be cut-out from this Gecko material.

192

Contact
Luxia
+56 (0)2 895 5057
sbenaglia@luxia.cl
www.theanemix.com

194

The Anemix®

The Anemix® is a lighting system which creates unique 3D effects. Each panel consists of a luminescent layer and a reflective layer which can be combined to give numerous different visual effects.

Composition
PMMA (polymethylmethacrylate), mirror and LEDs

Properties
Decorative, deep-sea effects and numerous motifs possible

Applications
Furnishing (table platters, etc.), communication or signage aids, ambience lighting, architectural elements (wall, floor, ceiling), etc.

Technical specifications
· Six types of modules are available. They can be assembled to create large surfaces, vertical or horizontal
· The complete system and its elements can be adapted to specific requirements
· Lighting is provided by OSRAM LEDs

1

2

4

3

1 The Anemix® is modulated to compose an entire wall, floor or ceiling. As shown here with the ceiling of the bar at Club de Oficiales Quinchamali in Chile.

2 The hanging lamp Voromix is designed by gt2P in collaboration with The Anemix®.

3 The Anemix® material was used to decorate a wall at one of the offices of Limitless, a real estate company in Dubai.

4 Ministry of Economy, Santiago, Chile.

Contact

NanoSurfaces
via Matteotti 37
40057 Cadriano di Granarolo
Emilia (BO)
Italy
+39 05 1606 5900
info@nanosurfaces.it
www.nanosurfaces.it

TiColor

Composition
Titanium

Properties
Decorative, resistant and a large colour range

Applications
Interior and exterior architecture, panels, industrial products, medical appliances, etc.

Technical specifications
· Maximum dimensions: 800 x 800 x 400 mm
· The colouring technique can be used on ready-formed parts with a complex shape
· Motifs can be produced (stripes or rainbow, for example)

TiColor is an electrochemical process for titanium surfaces, allowing unusual, intense colours such as pink, blue, green, etc. to be produced. By a controlled increase of a few nanometres in the thin layer of naturally-produced oxide which protects the surface of the titanium, modification of the effects of light diffraction at the surface increases the colouring possibilities. These colours are chemically resistant, suitable for interior as well as outdoor use. When the colours are exposed to sunlight, they become particularly intense and dynamic.

See notions: **Nanoparticles** p.226

196

FRAME #52
Sep/Oct 2006

Contact
Fornace Polirone
Via Argine Boccadiganda 1
46030 Borgoforte (MN)
Italy
+39 03 7664 8612
info@polirone.com
www.polirone.com

Variegato

Composition
Fired clay

Properties
Decorative, strong and hand made

Applications
Exclusively for interior floor covering: individual houses, inns, restaurants, historic monuments, etc.
Use outdoors can possibly be envisaged if sheltered well away from bad weather, with a minimum temperature restriction of 4°C (39°F).

Technical specifications
· Available in several shapes (square, rectangular, hexagonal, diamond, etc.) and in several sizes up to 400 x 400 mm
· Can be made-to-measure or as reproduction of ancient models
· There are classic tiles (joints 3 to 4 mm) or rectified tiles (joints 1 to 2 mm)
· Dimensions are given with a normal tolerance of +/- 1.5%
· Finish: smooth, rustic or sanded rustic
· There is no guarantee against frost

Fired-clay floor tiles are made by manual pressing of clays of different colours (yellow, red, black) into wooden frames. The clays are mixed according to an ancient technique, creating astonishing colour effects. The effect obtained is rather like the coloured moiré end-papers used in old book bindings.

1

2

1 Villa Santa Caterina, Grazzano Badoglio, Asti, Italy.

2 Villa Porro, Lonate Pozzolo, Italy by architect Andrea Bollazzi.

198

Contact

Dr. Mirtsch
Ruhlsdorfer Strasse 95
14532 Stahnsdorf
Germany
+49 (0)3329 699 576
info@drmirtschgmbh.de
www.woelbstruktur.de

Vault Structuring

Composition

Aluminium, steel, cardboard and plastics

Properties

Gentle, bionic self organized process, very rigid and light, decorative, light play and heat conducting

Applications

Architecture and interior architecture, automobile, furnishings, electrical domestic appliances, packaging, light fittings, etc.

Technical specifications

- Aluminium sheets up to 1.2 mm thickness
- Steel or stainless steel sheets up to 1.0 mm thickness, width of coils or plates up to 1200 mm

This technique for forming hexagonal 3D shapes from thin material such as metal, plastics, paper and cardboard sheet increases the rigidity of the materials. Thus lighter parts can be used with no loss of rigidity. Furthermore, there is decreased sound reflection from sheets with this structure, heat transfer is facilitated and when light plays on the hexagonal motifs, there is no dazzling effect.

Contact
Marotte
47 rue Eugène Berthoud
93402 Saint-Ouen Cedex Paris
France
+33 (0)1 4948 1360
www.marotte.fr

Woven wood

Composition
Carved and varnished MDF, and wood from fine species of oak

Properties
Decorative and strong

Applications
Wall coverings

Technical specifications
· Dimensions: 2500 x 1220 mm
· Finish: ecological water-based satin varnish

This decorative panel of cabinetmaking quality in 'golden' oak has a stylish grained surface. The manufacturing technique combines two species of wood veneer glued perpendicularly to each other. The fine surface species is then partially sanded, reproducing the weave of a textile, where the contrasting lower layer is revealed.

210

Contact

Wovin Wall (HQ)
6-8 Ricketty Street
Mascot NSW 2020
Australia
+ 61 (0)2 9317 0222
esales@wovinwall.com
www.wovinwall.com

Wovin Wall System

1

This covering system is effected by clipping tiles of different types of materials (laminate, plastics, metal, etc.) to a light aluminium frame. The curvature of each tile, alternating with that of its neighbours, creates a rippling effect over the surfaced thus covered. Once the locating grid has been put up, it becomes very easy and quick to change the appearance of a wall or ceiling by simply changing the tiles. The system acts an acoustic absorber (this has been tested and certified) and can also be backlit to fully emphasise the 'movement' of the tiles.

1 In a busy corporate cafeteria area the clients were astounded by the impact installing the Wovin Wall panel had on quieting the sound in the space.

2 Lollipops have personalised their child care centres by printing photographs of the children who attend on the Wovin Wall Tiles.

3 The I-Square shopping centre thoroughfare was brought to life using stunning red Wovin Wall panels

Composition

Aluminium framework, tiles: polypropylene, laminate, metal laminate, timber laminate, aluminium (which can be anodised or powder coated), digital printed images on PETG or PVC, polypropylene screen printed to any PMS colour

Properties

Decorative, relief effect, light, numerous types possible, acoustic absorber, and easily and quickly put up

Applications

Wall or ceiling coverings

Technical specifications

· Two types of tile: standard and oval
· Aluminium mounting grid module: 240 x 240 mm
· Two types of framing: standard or deluxe

2

212

3

214

Zelfo

Contact
Omodo
Am Wasserturm 1
16247 Joachimsthal
Germany
+49 (0)33 3616 4931
info@omodo.org
www.omodo.org

The Green Factory
www.thegreenfactory.fr

Composition
Cellulose extracted from recycled (paper or fabrics) or rapidly renewable (hemp, jute, miscanthus…) sources.
Natural pigments can be added.

Properties
Decorative, structural, biodegradable and strong

Applications
Interior panels, construction, furniture, musical instruments, acoustic products, consumer products, transport, etc.

Technical specifications
A spectrum of colours is possible: black, white, green, blue, red, yellow, purple, brown, ochre, etc.

Zelfo Technology develops and optimizes cellulosic materials for the advancement of sustainable manufacturing. Introductions to industry partners for end product purchase and development, raw material fibre sales (dried product), technology transfer and licensing of Zelfo dedicated CORE (Cellulose Optimising Reduced Energy) fibre processing systems are among the missions of this company.
Zelfo is a beautiful, warm, patinated or single coloured material, 100% natural. It is entirely biodegradable and its manufacturing process is purely mechanical, requiring no complex chemical techniques.

1

1, 2 & 4 Several samples of the Zelfo material.

3 Designer Elise Gabriel has collaborated with The Green Factory, to design 'The Zelfo Embrace', a series of furniture pieces which are constructed using the material Zelfo. Here you can see the the Zelfo Embrace trestle.

3

2

4